TURNING WE

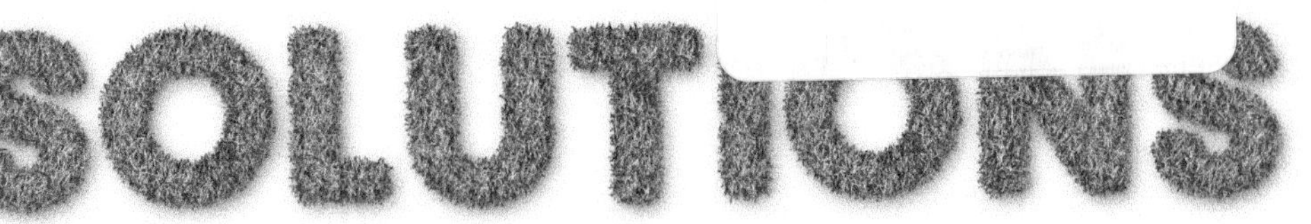

READ THE WEED

Global Publishing Group

Australia • New Zealand • Singapore • America • London

TURNING WEEDS INTO SOLUTIONS

READ THE WEED

Gwyn Jones

DISCLAIMER

The information and data in this publication is for **educational purposes** only and should not be taken as or a substitute for professional advice. Therefore, while the information contained in this publication has been formulated in good faith, the contents do not take into account all the factors, which need to be considered before putting that information into practice. Accordingly, no person should rely on anything contained herein as a substitute for specific professional advice.

Under Australian Federal and State laws some weeds are classed as prohibited / notifiable / restricted. All outbreaks of these weed classes must be reported to the appropriate control authority as they are considered a potentially serious threat to primary production or the environment.

For international readers in different countries, nations, states, counties, provinces etc, there are laws (weed regulators / agencies, legal precedents, cultural precedents and legal theory) associated with weed control, management, eradication and monitoring etc. Therefore, in regard to weed control, management, eradication and monitoring etc, all legal, environmental and cultural regulatory compliance and necessary governmental laws should be adhered to.

The material in this publication is not and should not be regarded as legal advice. Users should seek their own legal advice, especially in regard to weed regulatory agencies at the federal, state, local and cultural levels. Integrated Agri-Culture P/L and or the author will not be held liable for the misuse of the contents of this publication and reserves the right to update or eliminate them or establish limits or restrict access to them at any time on a temporary or permanent basis. If from this publication you require advice in relation to any legal, financial or business matters, consult an appropriate professional. In this publication specific registered products may sometimes be mentioned, but only as a guide to appropriate treatments. This should not be in anyway interpreted as a specific endorsement. Conclusions drawn from this information are the responsibility of the user. The user of this information assumes all liability for their dependence on it.

First Edition 2023

National Library of Australia
Cataloguing-in-Publication entry:

Turning Weeds Into Solutions: Read The Weed - Gwyn Jones

1st ed.
ISBN: 978-1-925370-84-3 (pbk.)

A catalogue record for this book is available from the National Library of Australia

Published by Global Publishing Group
PO Box 258, Banyo, QLD 4014 Australia
Email admin@globalpublishinggroup.com.au

For further information about orders:
Phone: +61 7 3267 0747

Further Information

There are plants called Weeds of National Significance or WoNS, please see: https://weeds.org.au

https://www.dcceew.gov.au/environment/invasive-species/weeds

Australian Weeds Strategy 2017–2027, 2017, p.4, Invasive Plants and Animals Committee, Commonwealth of Australia, Canberra.

https://www.aph.gov.au/Parliamentary_Business/Committees/Senate/Environment_and_Communications/Completed_inquiries/2004-07/invasivespecies/report/c05: 'Conclusion 5.152 The Committee believes that the management, funding, community understanding and political will to address the issue of invasive species already within Australia is fragmented and insufficient. Mr Edward McAlister, AO, the Chief Executive of the Royal Zoological Society of South Australia captured both the scale of the problem and the hope that it is not beyond us: The problem seems immense and there is certainly no 'silver bullet' for all, or perhaps even any, of these pest species, either animal or plant.... Accepting that the problem is immense and certainly widespread, there appears to be a number of things which can be done. (492: Mr Edward McAlister, Submission 75, p. 5.)'

I dedicate this book to an incredibly patient man, Geoff Leske of Timboon, Victoria, Australia. A fellow dairy farmer and without his persistence and guidance I would not have had the head start that I gained to learn to read the soil, plants and weeds on our dairy farm and then beyond into foreign lands. Geoff was my best man at our wedding and since then he has gone home to be with the Lord. Geoff made the time to mentor me, correct me and also let me make my own mistakes. His simple reminder of saying *"Don't forget the shovel"* has stayed with me as half the solution to reading the weed is understanding its root system. From Geoff to me and me to you - *"Don't forget the shovel!"*

Gwyn Jones

FREE SPECIAL OFFER

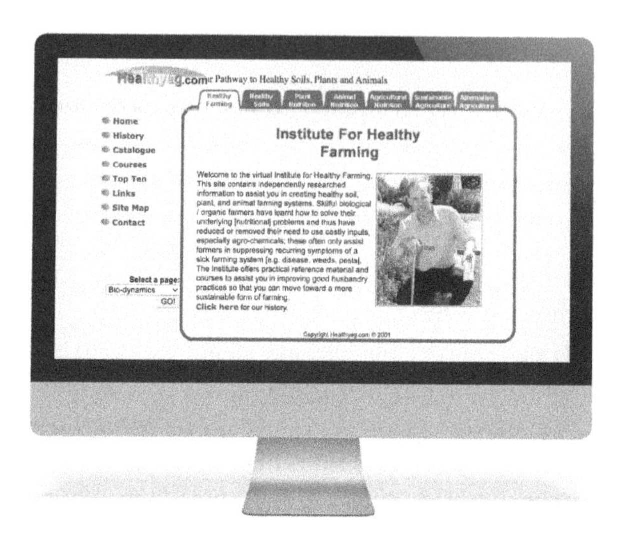

As a special offer for buying this book I am offering a free bonus chapter - Turning Weeds 'On' and 'Off.'

This new standalone chapter could not fit into this book and I did not want you to miss out.

Aim to get a head start as I take you to a whole new level in my next book - Reversing Climate Change Solutions: Water, Weeds and Why?

I don't want you to miss out on this free bonus offer.

Go to www.healthyag.com

CONTENTS

"Weeds are the plants of nature's choice, whether we choose to culturally accept them or not."

Gwyn Jones

GradDip Sust Ag [UNE], MRur Mgmt Studs [Syd]

FOREWORD

An Original Book on Weeds

This is an important and original contribution to our understanding of how landscapes function. It is built around the key insight that weeds are the symptoms of either landscape change or poor management – and not the problem in itself. The revealing use of a biomimicry model by Jones proposes that weeds are significant tools for landscape managers as important indicators of the real problem and/or state of health of their farms. That is, we farmers have been blaming the wrong thing: when the key issue is inappropriate landscape management. This clearly includes our failure to 'read' what so-called demonized weeds are usefully and diagnostically telling us concerning the health and/or ill-health of landscapes. The result, as Jones says, is 'the ongoing and accumulative decline in the Australian landscape."

This important book and its pertinent insights thus provides a significant tool for all farmers and landscape managers to both challenge our learnt paradigms and to better 'read' anew our degrading or regenerating landscapes. It thus provides a key pathway to regenerative paradigm practice change via new understandings and perceptions. As Gwyn Jones says: "Read the plants first, and not the herbicide labels," so as to learn "to work with nature and interpret nature's adaptive repair processes."

Charles Massy: Farmer and author of 'Call of the Reed-Warbler: A New Agriculture, a New Earth.'

"Read nature, not books!"

Professor William Albrecht

(1888 - 1974)

INTRODUCTION

Firstly, thank you for starting this weedy book as it was forty plus years in the making. I have often been described as 'different' as I think differently. I have a different view of this world, I am a big picture thinker and enjoy solving problems. As you will soon see, weeds have personally fascinated me from even from a young age and yes, some weeds I do not like such as The Stinging Tree or Suicide Plant (Dendrocnide moroides) as its stings are ten times worse than a stinging nettle and all the stings go off at once in reaction to cold. They literally give you agony, how do I know this, it took over a year for all the stings to disappear out of my leg! As this book will have an international readership, I have used the word – field and not the Australian term paddock and chapter six can also be related to European readers as many of these plants have originated from there. I have studied weeds in parts of North America, Europe, Middle East, Asia, Africa and seen similar weeds to that which are in Australia. My best memory to reading a weed form and function was in Nairobi National Game Park with ostriches, zebras and giraffes etc. On the flats was a relatively newly formed drain with its associated bare soil. Guess what 'exotic' weed had made its home there? What weed had been eco-triggered to grow? It was what I describe as a Scotch Thistle with its purple flowers. The very weed that changed my understanding of what a weed really is.

The context of this book involves the study of nature and what most of humanity is not understanding about plants called weeds. If you are working with nature and are involved in gardening, lawns, horticulture, cropping, farming, agriculture, landscape management, wildlife conservation or working with plants in general, these processes are all about understanding and working with plant succession. In the bigger picture, weed control and management is trying to control and manage a self-perpetuating, self-organizing, accumulative process called plant succession that gives life to this planet.

I am going to introduce you to some new terms and concepts such as the **plant successional staircase** and the term **Desired Plant Landing** (DPL), which is the best location or landing a position for your desired plants in the plant succession order. If your desired plants are not positioned in that landing you will have highly competitive weeds that are **eco-triggered** (ecologically triggered by a change in the environment) and use their specific **positive feedback loops** to repair a degraded plant succession and associated degraded landscape. As I have intensively studied and researched the topic of weeds, I quickly found out that I did not want to call myself a '*weedologist*' for that is quite a different topic! However, I did find out that regarding the scientific understanding about weed control, management and elimination, the 'elephant in the room' is the ignoring of plant succession. Do some homework yourself and find some weed control or weed management books, field guides or government policies and do a word search on plant succession. Often there is no linking of weed control and plant succession – Why? That is the reason why about one third of the book involves plant succession and plants lists, one third on the functionality of weeds and meeting the large 'Weedy' Family, yes, all 17 of them! One third of the book and what I start with, is about my journey on understanding the nature of weeds and how and why this book got written for you to learn how to turn weeds into solutions. Please remember "*Nature does not have weeds; weeds are a human innovation*."

CHAPTER 1

WHERE DID YOUR UNDERSTANDING OF WEEDS COME FROM?

CHAPTER 1

WHERE DID YOUR UNDERSTANDING OF WEEDS COME FROM?

I started out life seeing the benefit of weeds, but then I learnt to hate them and spray them, which did not work. Then blisters turned to calluses as I continued to wage my war on weeds until one mistaken act happened and then I realised the value of a weed. This is one of the reasons I decided to write this book. As a young person in Australia, I would help my father in our home vegetable garden. My parents were migrants and they brought with them their English culture. In England they were both involved in the war effort of World War II as my father was a Royal Marine and my mother worked in an aircraft factory. During the War they both experienced food rationing, so they highly valued the limited food supply. After the war they had an 'allotment,' which was a very small piece of land that they did not own, but were able to grow food on it. On that small plot of land, they had to maximise its production, yet they could afford only a few inputs to go into it. I give this as a background as they brought with them and continued their gardening culture in Victoria, Australia. Over the cold winter weeds would be allowed to take over most parts of the vegetable garden and then in early spring we would start 'trenching' which was manually doing what an old-fashioned plough used to do. A trench was dug the width and depth of a shovel and the weeds from the next area to be dug were chopped off at the base and dropped into the trench. Then the area cleared of weeds, that was a shovel width across, was dug

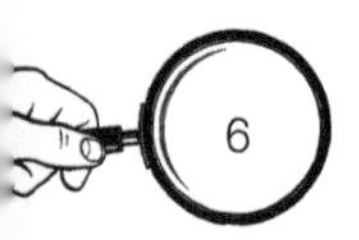

out and placed on top of the weeds. This 'trenching' process was repeated again and again. The weeds were the starting fertiliser and the secret was not to dig too deeply into the clay. The process was the same as a plough that would just take off about an inch (25 mm) of clay and lay it sideways back into the soil, especially when it had skimmers that dropped the green trash into the bottom of the ploughed farrow.

I was brought up thinking of weeds as plants that were going to feed earthworms and improve the soil for the coming season's crop. We used compost and any other green material into the trenches. Additionally, we used a light coating of lime added on to the soil surface, if there was too much clay. Beds were formed up and planted in. Whether I realised it or not, I was the son of a European family and I was being trained in a form of 2,000 year old 'Roman' agricultural soil improvement methods and culture.

May I ask you a question?

How has your cultural background, traditions and up bringing influenced the way you think about plants that are called weeds?

What is a herb? Is it a weed or a plant to be valued?

Why do some cultures eat and enjoy, what other cultures call weeds?

Take a moment and think back to when you first heard of the word – weed?

What was it in context to?

Was it associated with a plant being a problem?

I had a change of mind set or paradigm shift about plants as weeds when my parents bought a 'Walk in - Walk out' dairy farm, this meant that all the stock and machinery went with the sale of the property. This coastal high rainfall dairy farm was part of the Heytesbury Settlement Scheme near Simpson in the Western District of Victoria, Australia. On the farm's flatter low lying areas that were drained wetlands with black soils, we had what was locally called '*Scotch*' Thistles with purple flowers. We had a lot of big 'Scotch' Thistles and these plants were not what we wanted as they were taking up valuable space that we wanted for pasture. When I was about fourteen our family invested in buying a new back pack spray unit and with it I started my war on this weed using herbicide. I was skinny and the back pack was heavy and I quickly learnt to first lighten the load of herbicide on the biggest weeds. However, I noticed that where I had killed each big thistle there was a big dead spot and lots of small thistles were coming up in that newly created environment which I had created. So, we changed my strategy from an industrial (chemical) farming approach to simply hand hoeing (physical) using an adze, not a conventional chipping hoe.

In hindsight, I had not yet worked out why we had so many thistles in the first place. I thought that the weed was the problem and not the symptom of over grazing and opening up the sward creating bare spots that triggered thistles to grow. Using an adze, which is normally used for shaping wood, as I could go deeper into the ground to kill the tap rooted plant may sound simple, but if it was not done properly the plant would grow back. To kill this thistle, you had to go deep next to the thistle with the aim to cut through the earth and into the purple part of the large tap root. I would then rotate my wrist and pull the cut root and the plant out of the ground. The thistle's deep tap root enabled it to grow well into the summer

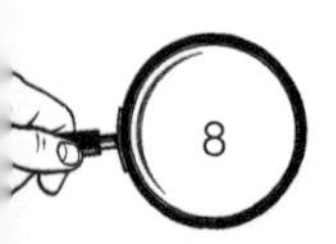

and it made its own '**biosphere**' or living home that covered and shaded the soil around it. As I dug out each thistle, I was amazed how many earthworms we had in our fields. I knew they were good for the soil as Charles Darwin wrote his last book about them back in 1883. One day I had to chop out some poisonous Ragwort by the roots and on lifting the ground out, I noticed how tight the ground was and that there were no earthworms. Being curious I started chopping into 'non thistle' pasture only to find no earthworms anywhere, I was shocked! Then I went back to chopping thistles and found earthworms and good well aggregated soil around their root systems. How come my pasture had no earthworms but my main problem weed was the breeding ground for them? Something was very wrong; was I wrong or was nature right? I had a major change of thinking or paradigm shift when I realised that the one plant species that was literally growing earthworms and topsoil was the one plant that I wanted to kill.

Over time the common weeds like 'Flatweed' (Catsear or False Dandelion), Ragwort and poor quality grasses disappeared as we transitioned away from the traditional fertiliser use of 2 parts Superphosphate and 1 part Potassium Chloride in spring and straight superphosphate in autumn as fertiliser to using mineral fertilisers including lime on heavy clay soils and the strategic use of trace elements. By strategically using lime in a low calcium soil, we made the phosphorous more available and did not need to add it. Looking back, we were in a high rainfall region with historically highly leached acid soils. On the soil surface was 'thatch' or poorly decomposed organic matter and the use of lime and specific trace elements started the creation of topsoil, especially the use of manganese sulphate as we had 2-6 ppm of manganese on our soil tests, when a minimum of 20 ppm was recommended for those specific soil tests. Using customised fertiliser mixes, mineral rebalancing occurred which assisted in decomposing the organic thatch layer that we had inherited and where it was used the Ragwort stopped growing. In the case of Ragwort this 'accumulator' plant, or should I say weed, was doing a job and when its accumulating job was completed, it had worked out its own destruction (Warning, 1909).

Lower productivity plants like Flatweed, Ragwort and poor quality grasses prepared the way for more highly productive plants in the progressive plant sequence. Another change of thinking occurred when I started to understand the difference from weeds being the 'Problem' to weeds really being a 'Symptom.' This happened when I started to value the relationship between plant species and soil fertility, soil biology and nutrient cycling. That is why over the years we literally out grew our weed problem by raising the overall soil fertility and associated plant succession. My major breakthrough of this overall concept came after reading about the success in New Zealand of de Faur (1966) where he classified fertility-demanding requirements to pasture plants and put them in a sequence. Below is the rising sequence from low fertility species to high fertility demanding species: Brown-top bent grass, Sweet Vernal, Crested Dogstail, Yorkshire Fog, Cocksfoot, Red Clover, White Clover, Paspalum, Timothy, Prairie Grass and Perennial Ryegrass. In Australia we do not have the bumble bee, so in general Red Clover does not last, but the other plants I could generally relate to in our cooler high rainfall climate.

I took an increasing interest in matching plants and soils. On the farm I looked at the different fields and related them to initially soil

type and then to soil fertility as our overall soil health improved. I saw distinct patterns in the different plant communities and communities within communities. I gained visual pictures of what plants grew in what environments and surroundings including the good and bad effects of grazing pressure and rest periods. As the soil types changed, so did the plant communities. Then I asked myself why and tried to work out the reasons for these relationships. I knew that by letting the grass grow out the tap rooted weed, described as Flatweed, was out shaded, especially by productive grass growth. Major hoof impact could set back bent grass and that was one reason why sheep farmers had more problems with it. I could see the Perennial Ryegrass, which was a high succession plant, close to the gateways where the cows would gather, there was lots of manure and associated high fertility and compaction. Yet at the other end of the field, there was a monoculture of Yorkshire Fog grass, which was a lower successional plant, because it had a shallower root system and it matured early becoming hairy and less palatable. As the fields were so large, we could not get enough grazing pressure on the back of the fields. Once we fenced each of the problem fields in half the Yorkshire Fog grass quickly disappeared, as we could use trampling and get more nutrient cycling and transfer to the back of the fields.

The following is written in my new Jones Phonetic Alphabet.

I waos verye glad too leave high school at fifteen and doo myi dairye farm auppreticseship. I enjoyed theu knowleidgle that school could give me, but I was verye bad with Einglish. I waos dyis (bad) lexic (woerds) and did not choosze too read myi fierst novel till I waos twentye owune. In myi mid forties I hated Einglish so much that I rewrote it byi breaking theu historical Einglish code. It took me nine yeiars too doo that and manuallye decode 12,000 main woerds. I have another book drarfted caorlled 'Dyslexic English - Not People: Real Solutions To Reading Challenges' wheaire I will jouerneye you throough whyi soume ofv us wholistic right brain dominaunt individuals have not woerked out that written Einglish isz an intentishonal code so that you have too get soumeowune else too teach it too you and that isz called theu educatishonal syistem. Byi breaking theu code so that everye individual sound (phfoneme) can be read in theu dyislexic Einglish code. This givesz newoo opportunities for aorll readers too upskill and too enhancse theair reading aubilitye, but that isz anouther storrye.

During my apprenticeship my wholistic right brain natural approach to farming was constantly being confronted by what we were being taught at trade school. The focus was to make more milk, we just had to add more nitrogen for more grass (and stop natural clover production). We were taught the need to use more nitrogen fertilisers, raise stocking rates and feed more grain in the dairy (milk parlour). We had to put more on and more on, until we became mor-ons! Looking back, I was being introduced to industrialised agriculture as described by Paul Newell in his concept of; Stewardship of self-organising land and water systems (please see Landmanship .com). I also listened to what the other agricultural apprentices were

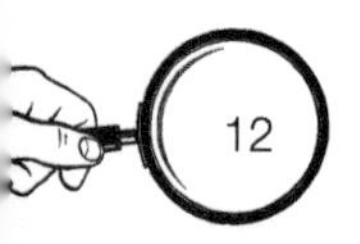

talking about including all the problems at work and weed control was a big issue. They applied nitrogen and sprayed the resulting nitrate weeds. They were very busy on tractors and could name all sorts of herbicides. Even back then I was learning to read the plants first, and not the back of herbicide drums. As an apprentice the simple message to me was maximum milk production, means more problems and more cost. It was like the dairy farmers were the ones who were getting milked of profit in the dairy industry.

I also mention this as a background that changed my farming ideas and my views of looking at all plants and not being too quick to give a plant a weed label and a death certificate. This process of wanting to understand more about working with nature and organic farming led me to a strategic and life changing event of meeting Lady Eve Balfour in England. My meeting with her occurred when my mother wanted to go back to England to see her aging mother. When the cows were dried off and not milked, prior to calving, we both went to England to see my grandmother. While there I contacted Lady Balfour who was the joint founder of what is commonly known as the Organic movement and is historically referred to as the (Sir Albert) Howard / Balfour Organic movement which included the British Soil Association. After contacting Lady Eve Balfour, my mother and I were invited to her home for a cup of tea. I listened to the way she thought and her constant use of the word – sustainability.

Thirty plus years ago this was not commonly used and the Organic movement was thought of as a group of weirdos. As my conversation with Lady Eve came to an end, she said that she was giving me a baton and I was to pass it on. My interpretation of what she meant was that she was very old and I was very young. I have literally taken on that task and it is a major reason why you are reading this book.

May I be so bold and also ask you to hand on the baton and share with others what you have learnt from nature as well as this book. Just before mum and I left Lady Eve's home, she suggested that we went to London to the Whole Foods Store as they had some books that would interest me. In brief I went there and asked to buy their whole library as each book had a price in it. When they realised it was going to plant new seeds of knowledge in Australia, they paid for the postage. The many books cost me a lot of money as I was only on an apprentice's wage, but one book, which cost 75 pence was called – The Weed Problem: A New Approach by King (1951) and the foreword was written by Lady Howard, the wife of Sir Albert Howard. She wrote: "*Mr King links the subject of 'controlled' but not 'eradicated' weeds with his well known no Digging*." King says that if we could ever banish weeds completely, then plants would also disappear. I did not realise the implication of those words until I gained an understanding of plant succession and how weeds were nature's first responders (Rogers, 2020) when the plant sequence was disrupted. Human disturbance and interaction for the purposes of exploitation creates the need for weeds to repair the disruption to the plant succession.

By buying the Wholefoods library, I started out with a large collection of post World War II books, which is the foundation of my large book collection that I have added to over the years. They all themed the same message of the future dangers of agriculture being industrialised, explosives were being turned into fertilisers and chemicals designed to kill humans, were being modified to kill insects. These books also raised the importance of natural or organic farming methods that were rapidly being lost and over taken by 'modern' exploitive farming and the greater ability to quickly kill large areas of weeds in one hit. Often the attitude expressed about weeds in these books was to learn from them and work with them where possible. A classic was Guardians of the Soil by Cocannouer (1950) and is a must read. In that era there was public discussion about soil, soil erosion and the need to slow it down. Books like Soil and Civilisation by Elyne Mitchell (1946) linked the destruction of soil to the destruction of future civilisations. Our modern civilisation is rapidly running into natural resource depletion. We are running out of topsoil and I am going to suggest that plants called weeds are the initial starting point for the regeneration of that process. Could it be possible that weeds are part of the solution to repairing the earth's degrading landscapes?

ACTION PLAN

Consider putting the book to one side and have a go and create your simple plant sequence list, the more plants the better. It will benefit you as we progress together in turning weeds into solutions.

May I ask you a question?

Have you thought about creating a local plant succession and what plants would be in it?

A simple plant succession is rocks, lichens, mosses, herbs, annuals grasses, legumes / clovers, perennial grasses, bushes, shrub and trees.

Could you make a short list of 5 or 6 plants and sequence them in an improving or productive order?

In shaded areas on rocks, stones or trees, do you have any lichens or mosses as they have no roots?

What is the highest successional plant in your landscape, garden, lawn, pasture, crop or orchard etc?

Is your desired plant the highest successional plant?

If not, why not?

Longer lived (perennial) plants are higher in a plant succession than shorter lived (annual) plants.

Do you have a legume that tends to be a higher order plant?

Now think which plants have tap roots or fibrous roots and why?

CHAPTER 2

ARE WEEDS A SYMPTOM OF AN ENVIRONMENTAL PROBLEM?

Land Management

"Degraded or disturbed land is known to be far more susceptible to weed invasion. For this reason weed control cannot be viewed in isolation from other land management practices."

Northern Territory Weed Management Handbook

2021, p 7.

CHAPTER 2

ARE WEEDS A SYMPTOM OR AN ENVIRONMENTAL PROBLEM?

The title for this book initially came out of a learning experience in a field when I realised that the weed that I was wanting to kill was the only plant group that was regenerating our soil and growing topsoil. I learnt that by reading the 'whole' weed, I realised I could start turning weeds into solutions. By making the time to observe the form and function of a plant called a weed, you can gain an understanding of why it is there, and therefore insight on how to manage or, even better still, how to prevent it. I am repeating this important message again. A practical tip is that to read the weed, you need to **read the whole plant** and that includes its root system. When you just look at the tops of weeds for a solution often you are missing a major reason why the weed is there.

On our high rainfall coastal dairy farm, I learnt by observation that the weed that I was trying to kill created the habitat and **biosphere** (living home) for the earthworms. Their castings were literally daily producing topsoil around its root system and under the protection of its surface leaves. Later I fully realised and understood that the plant I viewed as a weed was involved in producing more 'fertility' than we could afford in buying in commercial fertiliser and with no transport cost or my time spreading it! Fertiliser went on the soil's surface, earthworms put it where it was needed. A thistle was no longer a weed to be killed, but a plant to be **eco-logically** (logical ecology) managed. The eco-logical approach to managing this plant was to raise the plant succession, so that the plants that I wanted were in some form of balance (homeostatic) with the existing environment and there was no place (niche) for them to germinate and flourish. When (native) plant communities or single crops are the most competitive plants in that environment or habitat, they out compete the plants which are called weeds. In brief, weeds are plants that we do not want or value in our human culture.

During my dairy farm apprenticeship, our group was told that a weed was a plant out of place. It was in the wrong place so we had to kill it. We had to learn all about plants, especially weed identification, so that we could work out what plant name to look for on the back of the herbicide drum. How I discovered that the concept of a weed was a plant out of place was wrong, occurred when as part of my apprenticeship I had to do a plant identification list and present a collection of dried plants and weeds. This is relatively easy if you collect plants in the growing season when there are lots of seed heads for identification, but I had to make the collection in a low growth period. The end result was that I had to go to another district and get some plant species from there. This got me thinking, why do some plants grow in some regions, but not in others. Surely the air borne weed seeds move around all districts, trucks accidently move them, floods move them and big dust storms move weed seeds interstate and even overseas. Then I heard a farmer say that when he bought in hay, he knew he would not have weed problems from it. I was the person who always asked lots of questions, so I asked him how come? Then he gave me that look, you know the one that says, "*You've got a lot to learn*!" His answer greatly impacted my thinking as he said that he bought hay from properties that had alkaline soils, which had alkaline loving plants. His property, like ours, was acidic and the alkaline loving weed seeds would not grow on his acidic soil. If this logic was correct, a plant cannot grow out of place as the environment will not support it. Plants grow best in their preferred environmental conditions. A plant or a weed can

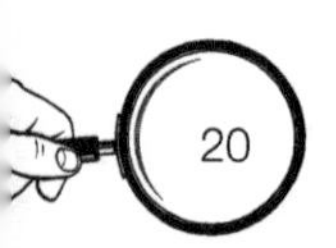

only grow in conditions and an environment that is conducive to its growth. The weed has to match the environment, not the other way around. Yes, weeds grow in pavements, driveways, on the sides of road or even in guttering where water accumulates. Human culture does not want the plant called a weed there, but nature does as that plant is restarting / regenerating / restoring a plant succession. I started to think about where else could I see weeds that were not out of place. As I rethought this question, I realised that when plants were sick looking, due to the environment not suiting their growth requirements, the stressed plants would improve if the conditions that favoured them also improved, but if the conditions worsened then they would often die. On the farm and in the garden, I had seen the plants that I wanted looking sick, but tap rooted weeds looked very healthy.

To say that weeds were a plant out of place was a human idea and theory, but in nature it made no sense at all. Plants called weeds grow exactly in the 'right' place as they are '**eco-triggered**' (ecological trigger) to spontaneously grow in an environment that suited them. In different environments, different weeds would voluntarily and naturally appear and healthily grow? If this concept was logical, if I wanted to control a weed then I had to control the environment. Change the environment, change the weed.

The problem was to decide between focusing on the environment or focusing on the weed. Which was more important? The more I looked the more I saw, the more I saw the more I learnt and this was not coming out of a text book. This aligns with the quote from William Albrecht, "*Study Nature, Not Books*."

As I started to focus on weeds and the reason of why they were there, many examples became obvious and the simplest was our farm house lawn. If I cut our lawn too close to the ground or cut it once and then quickly recut the new growth from the root reserve, thereby weakening the grasses root system, the grasses would be weakened and the lawn would open up allowing broad leaf weeds to establish in a newly created niche. Nowadays I do not focus on the broad leaves of a weed, I read the form and function of the weed, including below ground. What I know is that the new plant called a weed often has a tap root that can out compete the fibrous roots of grasses when the grass is in a weakened condition, often due to excess leaf removal. My eco-logical response to that was to simply raise the mower higher for a period and let the grass and its root storage recover. In the turf industry it is common to renovate an area with coring, which is the same function that a tap rooted plant does. In general, healthy lawns do not need coring and do not have tap rooted plants called weeds.

If the theory that specific plants or weeds only grow in a favourable or suitable environment to them then I asked myself the question, why is it that some weeds invade and produce a lot of seed and then they are literally gone? Have you heard the term – "*It's a bad season for XYZ weed*!" The seasonal conditions aligned with the preferred environment for that particular weed and 1,000,000's of dormant seeds were eco-triggered to grow. When the environment changes again the problem weed disappears as fast as it came. Warming (1909, p.348) reminds us that plant communities and their seeds are crying out as it were - '*Situation Wanted*' as million and millions of seeds are waiting for their right conditions. In regard to the cultural importance of killing weeds, so that they do not go to seed, if there are isolated problem weeds, it is culturally considered that the less weed seed going into the soil the better and this makes common sense. Yet consider the following, if in a given area there were 1,000,000 weed seeds in the right conditions and environment,

how many would grow? The general answer would be simply lots. If in a given area there were 1,000,000 weed seeds in the wrong conditions and environment, how many would grow? The general answer would be simply very few. This is why I would encourage you to initially look at the weed as an indicator plant of a changing environment and learn what the weed's form and function is telling you, so that you can turn this knowledge using biomimicry against the weed and transition from control to preventing a problem weed. As years turned into decades, I have observed and noted the publicity and associated anti-weed product selling as a weed problem takes the headlines. I have also observed that two neighbours could both have the same seasonal weed problem. One could spend a lot of time and money on killing it out, often with a need for repeated applications of herbicide as it just keeps coming back. The other neighbour had learnt to accept that the weed was a temporary problem and it would resolve itself. After eighteen months (a few seasonal rain events) the problem weed had disappeared, whether it had a chemical treatment or being monitored and the decision was made to allow the weed to complete its role (with the reasons why being documented in their property weed management plan). This situation repeatedly occurs as nitrate weeds invade landscapes after an extended drought period. A general theme throughout this book will be that the plant called a weed is not the 'Problem' but the 'Symptom.' The weed is growing where the environmental conditions allow it to grow. The problem is that from a management perspective you have normally unintentionally created its ideal growing environment to stimulate or eco-trigger the weed that you are looking at. The weed is the symptom of the environment and therefore I will be placing an emphasis on the relationship between weeds and their related environment.

"You need to get
to the root of your
weed 'problem'."

Gwyn Jones

CHAPTER 3

DON'T FORGET THE SHOVEL!

CHAPTER 3

DON'T FORGET THE SHOVEL!

"*Don't forget the shovel.*" Those were the words of my Bio-dynamic mentor as we were about to go from the dairy (parlour) to go for a walk over the farm. Those words were to remind me that we could not go and look at the whole plant, unless we also looked at the plant's and the weed's root systems. This next chapter goes deeper into why and how I learnt to read the weed and some of the logic of how I, and you could, gain the skills of using weeds as environmental indicators.

As I previously mentioned my parents owned a small 186 acre dairy farm below Simpson, close to Timboon in the Western district of Victoria, Australia. Back then we had a grass fed, 100 cow self-replacing herd and cut our own hay and silage. How I came to meet Geoff was because at the meeting in England with Lady Eve Balfour. I asked her about Bio-dynamics and she suggested that I write to Alex Podolinsky who headed up the Bio-dynamic movement in Australia. Later she sent me a letter with Mr Podolinsky's address and I wrote to him. He posted to me all his separate lectures all in one group (Podolinsky, 1985), which was an honour as back then you normally received one lecture at a time and then Alex would visit your property and literally assess if he was going to (freely) work with you or not. In Alex's letter he also suggested that I contact a local Bio-dynamic dairy farmer called Geoff Leske. Back then I was in my early twenties and had completed my dairy farm apprenticeship. I was the young pup and he was the older dog. Without his patience

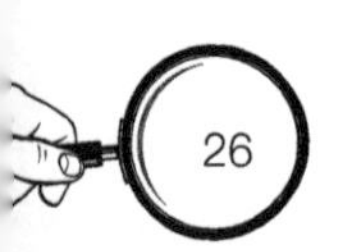

and more patience, I would not have been able to slow down my overly active brain to stop and look at things. Geoff laid the foundation of what is in this book by asking a few open questions that I have never forgotten. The questions that he asked me, you may also like to ask yourself the next time you are looking at some weeds as it is necessary to become more open minded and try and process what you see without your cultural bias. Look at a weed as a mirror of its surrounding environment, expressed through its form and function in order to gain an insight on how you can turn weeds into solutions.

On Geoff's first visit out in the field, he asked me a simple question – "***What do you see***?" Geoff knew what he was looking at, short over grazed pastures with shallow root systems. When he first asked me this, I just saw short green plants and a few that I could name. Then there were those 'other' plants whose names I had forgotten or wanted to forget. If I could not name a plant, they were not important so, to save embarrassment of not knowing them, I would call them weeds. If I called a plant a weed, Geoff would first smile and ask "***So what does it do***?" When Geoff first asked that question, I thought that it was an unfair question. I had a huge amount of respect for Geoff as Alex had suggested him and later we would talk on the phone for hours and hours, but to ask what does any weed do, did not make any sense back then. He was not talking about just one weed species, but all weeds in general. I knew about thistles and earthworms, but I decided not to tell Geoff as I thought, he may think I was a bit weird, so I just shrugged my shoulders. Geoff would smile again and with patience, rephrased the question, if I did not make any progress. "***What is its role, why is it there***?" Those broad type of questions would be the next part of my educational process. For me at that time my brain went straight to my limited knowledge of herbs. I had studied about herbs and herbs in pastures for I knew that many English authors like Frank Newman Turner (1951) had placed a great importance on them. They talked about herbaceous leys or something like that. The problem was that herbs were meant to be big and edible, like the stuff that got put on your plate at some fancy restaurant with an exotic salad or what was dried and used in a tasty meal, more my liking. So, my brain tried to link herbs and weeds and they did not go together. Then as I stopped thinking about what they looked like; for my weeds were small, coarse and often hairy as back then we consistently over grazed. Then I started to think about what they did, I started to re think Geoff's first question, which was the most important. What did I see? From Geoff's point of view, it was not just what was growing on top, but what they were doing underground as well. I should have realised this earlier as we were previously digging up soil and looking at it, this time, however, the focus was not on the soil but on the plants themselves.

As I connected what was happening above ground to what was happening below ground, I increasingly realised the connection between thistle roots and earthworms. As I previously mentioned, when we first started farming the only earthworms that we had was around thistles in the fields or in our vegetable garden. I had a love - hate relationship with thistles. I hated what was above the ground, because they just caused blisters and then calluses on my hands from hoeing, but I loved the thistle root system as it created topsoil for me in 6 -12 months! Geoff wanted me to think about the whole plant, roots and all. I thought that I was getting somewhere at last and as my self-confidence grew, I started to talk to Geoff about something I knew, which was thistles and earthworms. Geoff said that thistles preferred and were triggered by baring the ground and

if you create those conditions, then they will voluntarily arrive as that was their role. Then Geoff gave me that cheeky look of his! He did not have to say any words for I had worked it out for myself and I laughed and said "*So I used to spray to kill the weeds, baring the soil to grow more weeds. Then, things got worse and it had cost me time and money to do it*!" Geoff, being Geoff just smiled as if to say "*Lesson learnt*" and it was, but it was part of a bigger lesson. I was learning to work with nature and not against it. I had to learn as we were over grazing and not letting enough time for the plants and their root systems to recover. Voisin (1961) uses the term 'untoward acceleration' of grazing, which leads to reduced rest periods and failure of the system. Today Allan Savory's foundational book Holistic Management Handbook and the third edition: A Common sense - Revolution To Restore Our Environment, gives a more global overview.

Geoff as my mentor was my foundation stone, his guidance and patience taught me to slow down, stop and actually see what is in front of me. But sometimes actions speak louder than words. I remember the first time going to Geoff's house. As with most country places you always went to the back door, so you could take your boots off. There were a few steps to the door and on the right side of the door was a well-used garden fork. I naturally thought that it must have been for snake control. In Australia we get a lot of big highly venomous snakes that come up to the house and drink out of the dog's water bowl. I thought it was a bit strange to try and use a fork sideways as they would spring back on you. I knocked on the door and we had a cup of tea, had a chat and then we went out the back door again. As Geoff went past his hand automatically just grabbed the garden fork. Geoff always said "Never miss an opportunity to have a look (at your soil)." Geoff could read his soil and plants because he was daily looking at them. He was a 'true' Bio-dynamic farmer, as he thought that a garden fork opened up the soil a lot better in order to have a look at it. A fork damaged the soil less than a shovel and showed the soil structure better. When Geoff next came to our place, we would start talking about what we had not finished off previously. Then before we went to go down the field he would say "*Don't forget the shovel.*" On our place we had to use a long-handled shovel as our soil was so hard, yet Geoff had soft darker soil after many years of Bio-dynamic preparations and composting. For over 25 years I have been delivering on-farm courses and field days for a living and I still say "*I can't talk without a shovel.*"

In hindsight, what I did not know or understand was that my on farm Bio-dynamic training with Geoff, meeting a New Zealander and meeting Lady Eve Balfour in England, lead to Liz Clay and myself writing and delivering Australia's first commercial organic conversion course (16 days and partly federal government funded). During that course we identified that in dairy / beef farming after mastitis / calf scours, weeds were the next limiting factor to organic conversion. I then wrote and delivered 10 day (2x 5 days) courses on co-ordinated soil, plant and animal nutrition, while editing two books on Soil Health and Better Farming, gave 106 talks and lectures on soil health and carbon testing, followed by ten years learning and educating with Peter Andrews and his son Stuart of Natural Sequence Farming (NSF), which was a term coined by my friend Paul Newell. After twenty plus years of thinking about how I was going to address global, landscape and on-farm weed issues, the background was being moulded for me to write this weedy book and share how to turn weeds into solutions by understanding plant succession and that weeds are the repair plants that interface

between primary plant succession (historical forests and grasslands etc) and human disturbance causing wastelands and desertification.

May I ask you a question?

When is the last time you went to the effort of grabbing a shovel and looking under your weeds?

Even in a mulch garden or no till cropping, make the decision to look at a weed's root system.

Even if you only identify if it is either a tap or fibrous root system.

If 50% of the answers to solving your weed 'problem' was underground, just looking on top at best only gives you 50% of the answers. I would even go further to say that a weed's root system is more important to look at and understand than the top parts. The good news is that once you have done it a few times and linked your understanding of the top and bottom parts of the weed, you do not have to always dig up a weed, however, in cultivated ground it is often important to see if there are any hard pans and if the weeds are getting through it or not. I have often seen tap rooted weeds going through a hard pan and cereal (grass) roots failing to do so.

"…. a weed is the
right plant,
in the right place,
at the right time."

Tilley et al, 2022

CHAPTER 4

THE IMPORTANCE OF PLANT SUCCESSION

CHAPTER 4

THE IMPORTANCE OF PLANT SUCCESSION

If you have just opened up to this chapter and not read the previous chapters, you may be wondering what weeds have to do with plant succession and the simple answer is EVERYTHING. Weeds are the initial repair plants in a disrupted plant succession process being nature's first responders (Rogers, 2020). Rogers (2020) also suggested that "*weeds will proliferate as global warming and other human impacts intensify*." You may be thinking that you are fighting against weeds or that you just want to better manage your weed problem. However, I have some major news for you. When you are trying to fight against, kill or manage weeds, you are in fact trying to fight against or manage one of the earth's major energy accumulation production processes. You are like a very small tiny ant looking up at the front end of a huge roller and it is about to crush you. You are trying to come against plant succession, which is a self-organising, self-perpetuating, biomass accumulating process that propels the primary production of land-based ecosystems. It is like the earth's bio-power house that has helped sustain life on earth. In fact, without it there would be no earth as we understand it today, as plants (and weeds) are the earth's primary producers.

The theory of plant succession comes out of the theory of ecological succession whereby new species progressively supersede each other over time as they transition to a stable **end point or climax**. Each plant creates conditions that form beneficial circumstances that is repeated again and again with each sequence of

interdependent plants as the self-organising, accumulative plant succession develops. In the right conditions if you roll a snowball down a slope, it grows bigger and bigger. Plant succession has a similar 'snowballing' effect as it is a self-perpetuating accumulative process. It is like a super living organism as in the big picture concept of Clemence's (1916) 'superorganism' and Van Goethe (1790) concept of plant metamorphosis. If you put these two concepts together you end up with the metamorphosis of a superorganism that propels the primary production of land-based ecosystems. Yes, that is a bit deep and a mouthful to explain the perpetual function of plant succession. The above concepts are now considered to be out-dated and unscientific as wholistic concepts are dismembered in reductionist frameworks which dominate science. I acknowledge that in many ways we have academically moved to accept the 'normal' plant succession to include change and turmoil and that the process is being dominated by a 'recovering' from the last disturbance. However, I still relate to the overview of wholism and the phrase attributed to Aristotle – "***The whole being greater than the sum of the parts***." My old-fashioned views are more about systems, natural history, natural phenomena. biomimicry and natural based solutions, which are complex and dynamic being systems within systems. I read weeds, plant communities, livestock, soils, hydrology and landscapes, which is poorly documented due to inter-related complexity. What I have just said is that modern main stream rational left brain science does not align with my wholistic intuitive right brain thought processes.

Humanity is not yet understanding the full significance of systematically killing living organisms that supply our ability to survive on earth! The new global carbon marketing economy has been triggered by the climate change 'crisis,' but that is in part the symptom of a global problem being the depletion of mature plant successions through the human activities of agriculture, farming, grazing, cropping, urbanisation, industrialisation, mining, transport etc. Repeatedly exploiting plant succession robs the earth's natural regenerative ability to have accumulative response systems in order to sustain itself as primary plant succession transitions into secondary succession and degrades into wastelands and plant less deserts. As this has occurred the landscapes have released their retained former plant carbon into the air. Does it sound too simple to aim to reverse these processes?

May I ask you a question?

Do we need more plants on regenerating landscapes to utilise more carbon and not just focusing just on renewable energy and carbon budgeting?

Is it time not to focus on just producing less carbon, but also creating accumulative response systems through growing plants and regenerating more advanced plant successions? Could it be that with the best science available in the world, we are not winning the war on weeds as we do not understand the war we are choosing to fight?

To take a very extreme view, if humanity puts the resources into winning the war on weeds and pushes back the reparative ability of weed invasion, it would be putting to death the 'superorganism' that propels the primary production of land-based ecosystems in the form of plant succession. Thank you for your patience as I have digressed in part to explain why modern science is failing to find

the answer to the global weed problems, which is directly linked to plant succession, regenerative agriculture and the role of plants called weeds in resetting it.

How a Plant Succession Develops

As one plant grows by producing, accumulating and storing new cells to form its top growth and root system, dead cells accumulate as debris as they are dropped on to the soil's surface and also via root shedding. These dead cells decompose into a more concentrate form described as soil organic matter (SOM), which is like a warehouse of concentrated residues of the plant successional process. As more soil is formed then a food web develops, soil microbes increase and biodiversity increases. As the total mass of both, living cells accrue and the residues of dead cells accumulate the foundations are being laid, the upper environment builds to create improved living homes **(biospheres)**, that prepare and enable the transition to the next higher order plant's species to succeed and replace the previous plants. Each sequence of plants creates a more advantageous environment and conditions by utilising and building upon the decomposed residues of the previous plant community and soil microorganisms. In other words, each plant creates a more beneficial living home (biosphere) than the plant species before it and in doing so, it causes its own demise as the next bigger, taller, long lived plant replaces it. Each plant species enables its own demise as its ultimate death gives greater potential life to the new plant that overtakes it. This accumulative process continues until a successional climax establishes a stable dominant plant species and in Australia, for example; we have Red Gum forests, Blue Gum forests and Brigalow Country.

Plant succession theory is divided into two sections: **Primary** (original) and **secondary** (resetting) succession. Secondary successions happen after a major disturbance including the natural process like glaciers, volcanic activity, flooding, drought, wild fires and human activity like land clearing of a forest to introduce agriculture, cultivation of grasslands, including overgrazing. In theory a new secondary succession cannot go back to a primary one. The process of primary plant succession can start off with biological soil crusts (**biocrusts**) being formed when cyanobacteria and algae make their homes between very small soil / sand particles within millimetres of the soil's surface. This process is making the foundation for the next stage for lichens and mosses, which are rootless **cryptograms**. Small microhabitats are developed that create the micro climate and environment for small plants called forbs or herbs often with a tap root being able to anchor the upright plant. Annual forbs transition into perennial forbs and grow bigger as root–soil interphase improves. The next transition stage (**sere**) can include the inclusion of annual grasses with their fine root systems that will grow and focus where there are higher concentrations of water and or nutrient. Over time short-lived, then long-lived grasses progress to perennial grasses. If there is low rainfall (due to a rain shadow), grassland regions or prairies develop in the final climax as there is not enough rainfall to support trees.

To go into a bit more detail as the succession improves more diverse (spontaneous) vegetation develops that takes advantage of the improving conditions as plant residues, assisted by soil biology, form soil. More vegetation and residues reduce rain drop erosion, giving increased water infiltration and associated thermal regulation. These contributing components create improved environmental conditions as new biospheres are created. This in turn creates

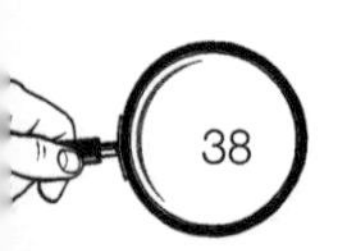

improved and more stable conditions for soil biology to increase in number and diversity, slowly forming humus (warehouse) as the end result of decomposition. As bigger and taller spontaneous vegetation develops, they become mini wind breaks that passively slow, trap and retain residues and slow the movement of water across the slope trapping other sediments. These processes increase nutrient cycling to enhance the microhabitats that accelerate with succession so that higher order plants can establish. Increased vegetation reduces surface runoff with greater litter layers increasing plant density and over time increased plant species diversity. These factors regulate and control raindrop impact, rainfall runoff and limit associated soil losses through erosion. As increasing amounts and diversities of vegetation residues, including root systems decomposing, more soil is made giving increased stability, resilience and equilibrium to the ecosystem. All the above effects create increasingly complex accumulative response systems that create a more sustainable biosphere and favourable environment for even taller voluntary vegetation to develop. As plants compete for light eventually canopies close in with mature higher order plant successional stages and a climax is achieved.

Having gone from bare ground to lichens, mosses, forbs / herbs to perennial grasses the next stage can be to legumes as atmospheric nitrogen is fixed into the soil. As carbon to nitrogen soil ratios become closer, higher quality decomposition occurs and more productive grass can take advantage. In theory the next major plant transition is to bushes and please note that in a degrading landscape the increase of bushes is a sign of increased desertification. Two main features of a bush are that they grow foliage to the ground and their branches have multiple tip endings (multi sprigged) with both features being natural defences against grazing animals. A bush is like a tent, protecting the soil and creasing a sheltered micro climate. As succession progresses the next stage (sere) can be scrubs and they have a single main stem and a vase shape, which can funnel water into the base of the tree. As each sere develops there are increases in nutrients, fertility and accumulating organic matter content. Topsoil depth increases and plants that are taller over shadow and dominate. The dominance of scrubs gives way to include annual then perennial trees, becoming more fungal dominant below ground communities, which contribute to a more stable complex climax forest and increased resilience. Forests draw in moisture and rain (biotic pump) and that is one reason why when forests are removed regional rainfalls can lower. This is important as rainfall is often associated with higher agricultural land prices and as water security and irrigation costs escalate, the value of maintaining natural water cycles is rapidly become of increasing ecological and commercial importance.

MUST KNOWS

Plants growing in a landscape are all in some form of sequence or succession. Most of what you are looking at is not an original (primary) succession, but rather a **human disturbed secondary plant succession**. In fact, it is now very hard to find and access a pristine 'untouched' primary succession of plants. It could also be argued that with a rapidly changing climate even the remaining pristine 'original' plant successions are under human disturbance and are changing.

As you look at plants growing in a landscape you are more than likely seeing a human disturbed landscape, ecosystem and plant succession. This is 'normal' and increasingly the term to describe them is a **'novel'** or new ecosystem. These human disturbed and managed ecosystems and landscapes dare I say, have weeds naturally in them as part of their biodiversity. To put it another way, you would expect to see weeds in a novel ecosystem as they are a characteristic of the ecosystem and may have initially re-developed it. This concept is very different from those who want to preserve conservation areas, wildlife reserves and national parks as foreign plants called weeds are not welcome in these pristine areas. As more primary forests are logged or cleared, primary wetlands drained and primary grasslands cultivated etc, there are increasingly less primary and more degrading secondary plant successions. It could also be argued that with the concept of climate change, all plants on the earth are affected by a global climatic disturbance, especially mountainous (conifer) plants. If global climate change is adversely affecting all plants, then does this mean that even 'primary' plant successions are now, to a greater or lesser effect, considered to be 'secondary' plant successions due to climatic disturbance? If this was the case, then it may in part help to explain why plants labelled as weeds are invading previously pristine landscapes. I will be taking this concept further in my next book: **Reversing Climate Change Solutions**: Water, Weeds and Why?

In summary; when you are trying to regenerate landscapes, build soil health and manage weeds, you are in fact trying to manage plant succession. This is why when past civilisations were failing, as the result of human disturbance, this caused a negative effect that led to degraded landscapes so that weeds moved in, it was only natural! Often after a civilisation failed then a newer higher order of plant community replaced the previous weeds that had completed their initial repair role. Alternatively, the weeds became naturalised in a (modern) novel landscape or the weeds failed to stabilise the ecosystem and the landscape collapsed becoming buried under the eroded soil and sand that the plants could no longer stabilise or maintain.

May I ask you a question?

Do you believe in climate change or a change of climate?

Does it really matter what name we choose to give or what theory we want to believe in?

If there has, or are, changes that trigger disturbances in plant succession, is the expectation of eradicating weeds, simply unrealistic?

Do we as individuals, or agencies, or policy writers need to think more about how we can work with weeds and learn to turn weeds into solutions?

CHAPTER 5

THE PLANT SUCCESSION STAIRCASE

"Weeds have evolved, for good and bad, to quickly invade and colonize sites opened by disturbance.

This is a trait restorationists should be exploiting rather than resisting."

Tilley et al, 2022

CHAPTER 5

THE PLANT SUCCESSION STAIRCASE

What I want to now explore is that plant succession can be bi-directional and globally we are seeing and experiencing a lot of degraded (permanent secondary) plant successions and highly dysfunctional, unhealthy landscapes. Plant succession is a self-organised sequential process that is generally slowly progressive, but it can also go very quickly backwards. As a result of a rapidly changing climate there is a need for an improved awareness that weeds are the new environmental sustainability gauges of climate change. Weeds increase as landscape degrades as they are the initial repair plants. Although, the previous chapter focused on the importance of plant succession and walked you through an accumulative (aggrading) plant sequence. In a primary succession if there is a disturbance the repair plants are called pioneering herbs, but in a secondary succession after a major disturbance these same pioneering repair plants are now called weeds, if there is human activity involved (Wilson, Jones, Paynter, Edser, Norris, Kravcik, 2023). Weeds come from the 'global plant list' and all weeds are plants, therefore all weeds are also in plant succession. Weeds follow the disrupting footsteps of humanity that is disturbing a progressive transitional plant succession. I mention the word **transitional** as the natural phenomena of plant succession is a self-organising progressive process until an end point succession or climax. However, from the limited life span of a human, it may take several life times for a major transition to occur in a slow growing ecosystem and associated plant succession. This is one reason why weeds do not only invade and start repairing the landscape, they can become naturalised as part of the new plant community for an extended period of time. Even to the extent of changing the plant succession direction into a future new novel ecosystem or become the new dominant naturalised plant, being often labelled as an environmental weed.

In this chapter I want to explore with you the modern-day reality that globally plant successions are going rapidly backwards or regressing, especially wetland species. This means that plant successions are bi-directional, they naturally and accumulatively, very slowly progress forward or can regress with increasing environmental stresses and very quickly go backwards due to human intervention (logging, overgrazing, cultivation, drainage or be eradicated with dammed water etc). There is a global process whereby primary forest, grasslands (includes Prairie, Tundra, and Savana) and wetlands etc are rapidly decreasing as degraded landscapes, wastelands and desertification is increasing causing ecological collapses and associated mass human migrations. This global transformation process is the degrading and destruction of higher order plant successions as humanity is quickly coming down what I am going to describe as a plant successional staircase.

I am going to use a **staircase** as an example (metaphor) to better explain the **bi-directional** changes and theoretical framework of plant succession as a staircase that you can walk up and down. A staircase can also have a landing, which is at the bottom and top of a flight of stairs or it can be in any part up the staircase. A landing is an extended level area (platform) that can also be used where the stairs can change direction. I am going to relate the top of the staircase landing to an end point plant succession climax as both are relatively stable and secure.

A climax is the conclusion of a 'mature' plant succession where there is with a steady stable plant population and living (**biotic**) environment within the limitations of a particular climate, environment and biogeography. A major feature is that after a sequence of species the last stage (**sere**) has a self-organised steady state or semi 'homeostatic equilibrium.' In other words, a state of dynamic balance resulting in a more stable plant community together with increased resilience and sustainability. Please remember that if you have a garden, turf lawn, horticulture, pasture, crop, orchard (especially with longer lived plants like trees) etc or if you are wanting to conserve a landscape (in time and space) and or its wildlife. In general, your aim is to maintain some form of stability / permanency / continuity / sustainable plant community that you are managing or monitoring. You are wanting to create a sustainable plant community in time and space, which I am going to describe as a '**Desired Plant Landing**' (DPL) area. The concept of a landing is an extended area (platform) in 'time and space.' The aim is to hold and maintain your desired plant (community) in 'time and space' within a naturally progressive plant successional order. In many ways the aim is to have your desired plants as the climax or end point of the sequence. Alternatively, the concept is to 'stretch' the length of time that you have for your desired plants to productively grow in a dynamic plant succession.

You want some form of stable healthy and productive plant community or monoculture without the plant succession going backwards (degrading) or progressing (aggrading), which **eco-triggers the weeds as repair plants**. It is very important to understand the role of having a Desired Plant Landing (DPL) within your regional plant succession (staircase), because if your desired plants are not matching their optimum to maximum environment conditions, weeds will fill in the successional 'gaps' on either side of your Desired Plant Landing (DPL). Below your DPL are often protective weeds as you transition towards an increasingly degraded landscape or above your DPL with potentially productive weeds and a trajectory change to a new novel 'weedy' plant succession.

How does your DPL relate to weed management?

The closer you can match your desired plants to your environmental conditions the more stable and resilient will be your Desired Plant Landing (DPL) area. By doing this there are less 'eco-triggers' to stimulate weeds to grow as your desired plants are highly competitive and can out compete your weeds. In general, weeds do not invade to transition and progress the plant succession if there is no repairing or upgrading of the plant succession that is needed. I have met landholders that have intuitively created their stable Desired Plant Landing area and it is as if the plant succession is standing still in time and space, due to their intuitive management. For me their most relevant comments are "*Well I don't do much as the place pretty much looks after itself*." Other comments can include: "*Nope, don't have many pests or diseases, I have no need for sprays*." Or "*Weeds just tell me I have stuffed up and I need to get back on track*." What does "*back on track*" mean? I would suggest that it is to rematch and improve the environmental conditions, to better match the needs of the desired plants. I believe these concepts are also drawing on the foundations of the Organic movement as the term "organic" farming was coined by Lord Northbourne, in his book, Look to the Land (1940) and he wrote that "*the farm as a living whole*." I would like for you to consider the concept of the whole; to include your own healthy successful Desired Plant Landing (DPL) area. What would it look like and what desirable plants would be growing in it? In your regional plant successional staircase,

where would it be placed? Imagine managing a plant community where there is not much to do and it could even appear that your management style is almost by neglect as the plants very much look after themselves. Imagine your self-organised plant community is part of a healthy self-perpetuating functioning landscape and in fact, your challenge could be not to do things that could interfere with its steady state. These outcomes would be learning to work with nature.

I have described a primary plant succession as a 'staircase.' The construction of a staircase normally starts from the ground and goes up one step at a time. One step is formed and that supports the formation of the next step. In the plant succession staircase each step represents a different landscape with higher order superseding plant communities. Similarly, in a plant succession, one plant forms the improved foundation for the next higher order plant, which builds upon the previous one. By using a staircase as a representation of a plant succession, each landscape step represents a sere or dominant plant community. Plant succession is an accumulative process that continues in a series of stages as the plants get progressively taller competing for light.

Now imagine you were standing at the base of the staircase of plant succession. As you looked up you can see each sequence starting with the ground level being with bare earth and rocks. As you progressively walk up the staircase, on the first landscape step there is just lichen, the next step moss with both having no roots. The next step is the tap rooted forbs (annual to perennial), the next step fibrous rooted grasses (annual to perennial), the next step nitrogen fixing legumes, then better perennial grasses, then ground covering bushes, then vase shaped shrubs, then trees, then forest being the end point landing called a climax. You are now at the top of the plant succession staircase and standing on a climax landing where it appeared to be stable as if time is relatively standing still. You have just walked up and seen an example of a primary plant succession.

An important part of primary plant succession is its accumulative underground component, which is generally unseen. This involves the slow accumulation of organic matter, formation of soil and soil biology creating topsoil. Now reimagine if you could look at a staircase on a side view, often under the staircase there can be an enclosed or covered in area. As each step goes higher there is a greater area underneath each step. The same can be said for the theory of plant succession, because as the plant succession progresses so does the quantity and quality of soil, soil biology and resulting topsoil formation. Humus is formed as plant residues decompose along with 'imported' insect and animal residues that originally came from the process of other organisms eating or taking shelter by the plant. Therefore, the plant successional staircase illustrates the progressive steps or stages of a superseding plant community and illustrates the **accumulative response systems** that increase the underground depth of soil development resulting from and jointly supporting the plants above.

I want to use the plant succession staircase concept to also introduce what is happening on our planet and how and where plants called weeds fit in. **To turn weeds into solutions**, you also need to understand where your problem weed or weed community is on the plant succession staircase and by reading weeds, we can also gain an understanding of which direction your landscape is heading (positively or negatively). **Weeds are nature's sign posts**

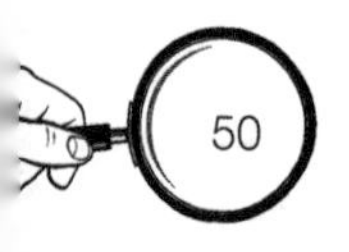

if we choose to look at them. Weeds can act like an environmental sustainability gauge and this is important in view of the ongoing effects of a changing climate. By understanding if your plant succession is transitioning towards and regenerating or backwards then you can make more informed management decisions on how to better control or manage weed populations. However, please remember that a high percentage of changes in weed successions are triggered by weather and especially water events being either too much or too little. Weather events are constantly changing which could mean that with another rain event it could change again and get better or worse. If you are considering more of an ecological approach to weed management then you have to increasingly start thinking not just when you have a weed problem, but well before it starts and understand why it is occurring or have you actually caused it.

The next activity I want to take you on is to introduce you to a downward or descending plant succession. As we take this journey, we will be directly involved in meeting nature's repair plants called weeds that are defending each landscape step and trying, as it were, to literally defend (protect) and hold the ground or soil that is around them. However, coming down the staircase there will often be only limited or few previous historical plants that you saw on the way up as the environment has adversely changed and degraded so much. The reason for this is that the changing environmental factors will have eliminated some of the previous plants and at the same time created (newly formed) degraded homes or niches that new plants called weeds will spontaneously grow in. These different plants are often foreign or exotic plants and can form a new '*foreign*' or novel ecosystem or landscape, as new environmental conditions need new plants as the pre-existing plants cannot match the changed environmental conditions. Coming down the staircase there will be an initial mixture of a few previous plants and an invasion of new foreign plants or weeds that are creating newly formed re-assembled plant communities. As you start to look down the now degraded plant succession staircase there is one consistent feature, all the trees have gone and looking three quarters of the way down there are only a few shrubs and below them some bushes. Further down the stairs you see less plants, more bare spots and loss of soil. It is a sad picture as it looks like what is being lost above ground is causing the loss below ground with eroding topsoil and subsoil being exposed. As you continue to look down the staircase it looks like plants are trading their ability to produce greenery in exchange for weaponry and finally to have the armoury to defend themselves. Are you are starting to realise, how highly degraded this global plant successional staircase really is?

Collecting your thoughts and refocusing as you stand on the stable climax landing, you glance down to the first landscape step below you, which looks so dark green as nitrogen fertilisers and excess raw compost has triggered fast growing nitrate weeds with their green soft foliage. It looks a bit of a jungle as these productive plants are growing so fast as the excess nitrogen has forced the cell walls to expand and weakened the external walls allowing swarms of insects to attack these weeds. The next landscape step down again is very different as it has cultivated areas and on some moist soils there are pioneer weeds quickly coming up to cover the disturbance of the new cultivation. The function of the pioneer weeds is to cover the soil, limiting wind and water erosion as it would start to remove the precious topsoil that the previous plant succession had accumulated. Below again are a few green, three leafed legumes in amongst some taller grasses. Legumes fix

atmospheric nitrogen, which closes carbon to nitrogen soil ratios and helps build humus. Then there is a bit of a shock as there appears to a higher percentage of more and more weeds on each of the remaining steps and each 'weedy' step down seems to go from **prickles to spines** to just a few **big thorny bushes** and bare spots everywhere. The next landscape step is different again with forbs, so you assume that they are mainly tap rooted weeds. The ground looks so tough that the weeds would need to have jack hammers to penetrate it! It also has weeds with weapons on them like spines and prickles, with even a few toxic ones. The next landscape step looks a lot drier with big thorny bushes and bare spots everywhere. There are a few shrubs, but they look under moisture stress. This next landscape step has grasses that look more like rusty wire and tussock or plants that just seem like mini fortresses. It was as though the plants all needed to have armour on to be able to survive. The next landscape step is very bare, but it has one feature, a lot of dig marks so the assumption could be made that there are bulbs or corms or gold nugget treasures stored underground by these plants and something is eating them. Are things so bad above ground that their only safety was to go underground and quickly grow in a good season? As you look further down, something is not right. At first you just think you are not focusing properly as you remember that previously you walked up each step. Then as you focus your stomach turns, you feel sick. No! It can't be, it must not be.

THE LAST 3 STEPS ARE MISSING!!!

You feel like there is something wrong as you remember walking up each step and they were all there! So why are they now not there! Why are you feeling sick and weak because you realise that you have just witnessed the demise of the earth's ability to biologically rebuild itself. The earth's mighty plant succession powerhouse or superorganism has died! You are looking at the void of nothingness, you are confronted with the realisation that things looked like they were getting progressively worse on the staircase.

What you have witnessed was a series of declining ecosystem services, overexploited to then becoming a disintegrating ecosystem, then overharvested to a denuded ecosystem leading to a collapsing ecosystem and finally the death (cessation) of an ecosystem with resource depletion and plant, animal (and human civilisation) extinction!

The previous visual story was meant to be confronting as you went up a primary plant succession order and then looked down at a rapidly degrading secondary plant succession that transitioned from a forested climax to an ecological collapse and wasteland. This confronting process is happening in many parts of the world as human disturbances such as urbanisation, industrialisation, transportation, cropping, agriculture and farming are transforming once primary climax plant communities into a degrading secondary plant succession. To turn this situation around a degraded plant succession needs to be regenerated and overall landscape functions enhanced through the creation of accumulative response systems. Could weeds be the initial solution to restart this regenerative process? I believe that they are!

May I ask you a question?

IF THE LAST 3 STEPS ARE MISSING!!!

How on earth are these last steps going to recover, if things look so bad?

Is there a point of no turning back whereby the weeds could not save the landscape, even the toughest ones?

What had caused all these plant losses?

What had destroyed all the wealth of accumulated topsoil and soil organic matter?

Have the loss and destruction of these higher order plant succession plant communities contributed to raising atmospheric CO_2 levels?

What was going to be the future fate of those that rely on dying landscapes?

"Thus, all farming and land management practices must take into account the succession processes involved, otherwise the control targeted at specific weeds might be ineffective or could even make the weed situation worse."

Scott (2000)

CHAPTER 6

INDICATOR PASTURE SPECIES IN SE AUSTRALIA

CHAPTER 6

INDICATOR PASTURE SPECIES IN SE AUSTRALIA

We have covered the importance of a self-organised plant succession, the **bi-directional** nature of using the **plant succession staircase** and the importance of creating your own **Desired Plant Landing (DPL)**. Now we are going to focus on a practice plant succession list. Globally there is a lot of documented ecological primary plant successional orders and yet the dynamic nature of degrading (agricultural) secondary successions and weed management appears to fall between the general scientific fields of botany, ecology, agro-ecology, agronomy, ecosystem restoration and natural resource management (to name a few). Is this due to academic authors wanting to avoid the 'hot' topic of adverse human intervention and the ethics of not wanting to take academic 'ownership' of what is often the study of a dying landscape? To me when it comes to scientific understanding about eco-logical weed control and management, the '*elephant in the room*' is the intentional destruction of global plant succession.

Having covered some of the natural complexity and academic challenges of documenting secondary successions, I want to put a practical and positive view point on a secondary agricultural plant succession by refocusing on my case study list. It involves a factual pasture sequence list for South Eastern Australia, which in part can be related to a temperate, cooler climate for international readers as well. This plant list initially came from plants on our family dairy farm having seen de Faur's (1966) New Zealand ascending list of

Brown-top bent, Sweet Vernal, Crested Dogstail, Yorkshire Fog, Cocksfoot, Red Clover, White Clover, Paspalum, Timothy, Prairie Grass and Perennial Ryegrass. On establishing my consulting company Integrated Agri-Culture P/L and becoming an independent agronomist, under the astute mentoring of Tony De Vere, I took a personal interest in pre-guessing what the soil tests were going to indicate by just looking at the plant species and then comparing what I thought, to what came back with the independent soil tests. Over time I got very good at correlating plant species to soil fertility and productivity. As I transitioned from just soil testing to independent (no product / commission sales) consulting. I quickly noticed that there were only about twenty general questions that clients asked, but in hundreds of different variations. This was the main reason why I first wrote and delivered my first four days; Science of Soils course which turned into the first Australian commercial organic conversion course over 16 days with Liz Clay.

As a consultant I would often intuitively answer questions about landscape, soil type, gardens, pasture, crops, weeds and interrow management through my mental understanding of soil fertility in relation to plant succession. Then after a few years one farmer said words to the effect, "*Gwyn, will you stop just talking about this succession thing and what you see in my paddock, just write it down so that I can read it and work it out myself!*" At the time I was a bit shocked as I was not intentionally withholding information, that is not my personality, it was more that I was a right brain visual wholistic thinker and had concepts in my head and he was a left brain sequential thinker and wanted to see it written down in a logical order. I thought like a vortex and he thought more like a ladder. So, I wrote a sequential plant list ladder for him. I still remember the first time I publicly showed my 56 Naturalised Indicator Pasture Species in SE Australia for Acid Soils List, when I put it up on the screen to a group of farmers and landholders. The hall went into silence, a long silence and as an educator I was thinking to myself this was going to be really good or ready bad. Then the room erupted with questions including two individuals having a major argument over the plant Phalaris. Phalaris is the plant's new name as previously it was called just Canary Grass and was originally considered a weed. I knew both individuals and one argued how good it was and how much money he had spent to get it going. The other person was saying how stupid that was as he hated the stuff and wanted to set fire to it to get rid of it. They both loved it. One loved it as his climax plant and the other person just loved to hate the 'weed' with a passion! The first person had mainly higher hill country that faced the afternoon sun and Phalaris with its deep root system could hold on and respond to the summer rains. The other person had lower flat country and had a higher order plant succession and thought that it was not palatable enough and did not fill his winter feed gap like Ryegrass did. So, in their own ways they were both right, but the first person had a lower stocking rate, non-breeding herd and made no hay or silage. For more North Eastern Australian land holders and managers the equivalent plant to Phalaris is Rhodes Grass. **Can you think of a similar plant in your region or in your nation?**

I am about to share with you my hierarchy of 56 Naturalised Indicator Pasture Species in SE Australia for Acid Soils. As you start to interpret the list it may appear to be very simple, but there is a lot of depth in the list such as a major two stage category for legumes and the major feature of excess productive weeds in the form of nitrate weeds. The significance of this is that I have referred to the need to raise plant succession on degraded landscapes so that the

desired plants out compete the (protective) weeds. However, there is a limitation to this concept and that is if you put more fertility and more on, you will become a more-on! High to excess nitrogen soil levels will eco-trigger nitrate weeds. Again, remembering that you are aiming to create your Desirable Plant Landing (DPL) on your own plant succession staircase and if you over fertilise or have lower soil biological (fungal) activity the productive nitrate weeds will be eco-triggered. The reason for this is that soil microbes eat at the first table. My message is that it is very possible to over fertilise, therefore, observe and document your own plant succession (or use the one in this chapter) to see how close you are to potentially over fertilising and then back off the fertiliser (and nutrient transfer) inputs and focus on building soil health and biology. Think about how you can become a better regenerative landholder with lower inputs. Also remember there is a stage beyond regenerative agriculture. The last stage is soil energisation, which I cover in my course, SOILCARE = **C**onservation, **A**ggregation, **R**egeneration and **E**nergisation (Jones,1999).

I use the term 'naturalised' for plant species that will self-propagate and therefore are self-sustaining. The list of naturalised indicator pasture species is not intended to be an exclusive list and deliberately does not include many "problem" weeds, so that the focus stays on the actual plant sequence.

The following comments are going to walk you through some of the plants on the list and explain their form and function. The aim of the list is to illustrate an ascending (secondary) plant succession, but I wanted to commence with historical native pastures as they were an important starting point. The commencement of what was described as pasture improvement would often have included the clearing of all taller vegetation and then the cultivation of the ground to form a rough seedbed. Superphosphate and essential trace elements, often including molybdenum, was included to stimulate legume growth. Lime was also added to low calcium soils (not pH), if the farmers could afford it and transport costs were low.

For the purpose of making the list, I have commenced with native pasture plant species and making the assumption that the soil is unimproved or has not had any fertilisers added. In general, all pastures have had historical native plant communities. In order to reset the plant succession on the list, imagine if the native pastures were destroyed (by Extreme Wildfire or Flood etc) or majorly disturbed by intentional cultivation and the sown 'improvement' pasture seed had failed to establish. The list shows the next group of plants would be the first Pioneer Volunteer Species in a low soil fertility class. The next succession group of plants to occur if this section was overgrazed and over mown would be plant communities like Bent Grass, Sorrel, Onion Weed, Flat weed and Ribwart Plantain. Onion Weed with a bulb and both Bent Grass and Sorrel grow with their thick subsurface (rhizome) roots have large storage root systems under ground. If over grazing / mowing occurs, these underground storage plants will not be easily killed, but could recover very quickly in improved conditions. Both Flat weed and Ribwart Plantain had leaves that, with limited competition, could be flat to the ground and difficult to be grazed by cattle or cut by a mower. They also had a large deep taproot that had the ability to access water and minerals, and then store it. As soil fertility and grazing methods improved the Medium Opportunist Plants established, which were often taller or more productive and could outgrow and shade out the lower fertility class. Again, as fertility improves and palatability also improves the plants in the high soil fertility class start to dominate

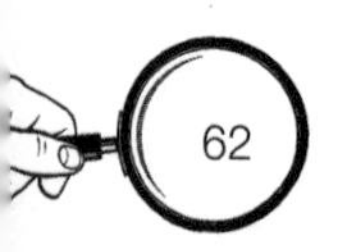

as mainly perennial plants give an increased growth and longer growing periods / seasons. If excess soil fertility occurs the plant succession degenerates with the invasion of nitrate accumulating plants to utilise the excess nutrients (Excess Nitrogen = Biology Deficiency). My understanding of secondary pasture succession initially came from New Zealand. My list is an extension de Faur's (1966) works and introduces a temperate Southern Eastern Australian pasture plant succession list, which in part illustrates a raising (often non-native) plant succession. The list is categorised by general soil fertility and introduces the 'production' weeds, due to excess nitrate and the repair plants called nitrate weeds.

There are five general soil fertility classes and related plants.

Soil Fertility Class: Unimproved	Native Species Lower Soil Phosphorous
Soil Fertility Class: LOW	Pioneer Volunteer Species Lower Soil Phosphorous
Soil Fertility Class: MEDIUM	Opportunist Species Moderate Soil Phosphorous
Soil Fertility Class: HIGH	Advanced Species High Soil Phosphorous
Soil Fertility Class: EXCESS N	Nitrate Weeds High Soil Nitrate

CORRELATION OF PLANT SUCCESSION AND SOIL FERTILITY

A Hierarchy of 56 Naturalised Indicator Pasture Species in SE Australia for Acid Soils

Soil Fertility Class:	**Unimproved**
Native Species:	**Original Plants**
Spear Grass	Stipa spp
Windmill Grass	Chloris truncata
Kangaroo Grass	Themeda spp
Red Grass	Bothriochloa macra
Wheat Grass	Elymus scaber
Weeping Grass	Microlaena stipoides
Wallaby Grass	Danthonia spp

(Then a major disturbance with initial land clearing)

Soil Fertility Class:	**LOW**
Pioneer Volunteer Species	**Repair Plants**
Mosses / Liverworts	Bryophytes spp
Toad Rush	Juncus bufonius
Winter Grass	Poa annua
Common Rush	Juncus polyanthemos
Spiny Rush	Juncus acutus
Sorrel (Sheep)	Rumex acetosella
Onion Weed/Guilford Grass	Romulea rosea
Trifolia and Medics	Trifolium glomeratum
Summer Grass	Digitaria sanguinalis
Meadow Foxtail Grass	Alopecurus pratensis
Soft Broome / Goose Grass	Bromus mollis
Flat Weed / Catsear	Hypochoeris radecarta
Bent Grass / Brown-top	Agrostis capillaris
Sweet Vernal Grass	Anthoxanthum odotatum
Crested Dogstail Grass	Cynosurus critatus
Ratstail Grass	Sporobolus capensis
Rat Tail / Slender Fescue	Vulpia bromoides
Broad Leaved Plantain	Plantago major
Ribwart Plantain	Plantago lancealata

The plants above are generally associated with lower soil phosphorous levels and the plant succession can benefit from raising the soil phosphorous levels.

The following plants in the MEDIUM: Soil Fertility Class are associated with moderate soil phosphorous levels and benefits may be gained by increasing the availability of the existing phosphorous in the soil ecosystem.

Soil Fertility Class:	**MEDIUM**
Opportunist Species	**Medium Production Plants**
Annual / Wimmera Ryegrass	Lolium rigidum
Subterranean Clovers	Trifolium subterraneum
Barnyard Grass	Echinochloa crus-galli
Tall Wheat Grass	Aropyron elongatum
Strawberry Clover	Trifolium fragiferum
Red Dock	Rumex pulcher
Swamp / Curled Dock	Rumex brownii / crispus
Yorkshire Fog Grass	Holcus lanatus
Tall Fescue	Festuca arundinacea

The following plants in the HIGH: Soil Fertility Class are associated with high to excess soil phosphorous levels.

Soil Fertility Class:	**HIGH**
Advanced Species	**High Production Plants**
Kikuyu	Pennistetum clandestinum
Phalaris	Phalaris tuberosa
Paspalum	Paspalum dilatatum
Prairie Grass	Bromus unioloides
Timothy Grass	Phleum pratense
Cocksfoot	Dactylis glomerata
Chicory	Cichorium intybus
Dandelions	Taraxacum officinale
White Clover	Trifolium repens
Perennial Ryegrass	Lolium perenne
Red Clover	Trifolium pratense
Lucerne	Medicago sativa

Soil Fertility Class:	**EXCESS**
Nitrate Weeds	**Excess Production Plants**
Barley Grass	Hordeum leporinm
Capeweed	Arctotheca calendula
Wire Weed / Knotgrass	Polygonum aviculare
Hog Weed	Polygonum patulum
Fat Hen / Goosefoot	Chenopodium album
Marsh Mellow	Melva parviflora
Prince of Wales Feather	Amaranthus refetroflexus
Black Nightshade	Solanum nigrum
Common Stinging Nettle*	Urtica dioica

* Silica accumulator and square stem.

Other nitrate plants can include: Corkscrew, Lamb's Quarter, Pigweed, Caltrop etc. High risk soils are shallow with high organic matter and especially after 'drought breaking rains', when there is a short flush or soft green feed, nitrate poisoning of stock can be unseen (subclinical) and trigger Grass Tetney and general ill health.

The above list in part illustrates a raising (often non-native) plant succession. The list is categorised by general soil fertility and excess fertility with excess nitrate and associated nitrate weeds. As an educational pioneer in organic farming, I was often asked to explain my understanding of how to improve productivity and I would inevitably relate it back to plant succession. Then we would go for a field walk and I would show them their regenerative landscape's plant succession by starting at their worst and lowest order field, then a mid-order field and finally what they considered as their best field. This journey would also be related to soil and soil audit results if available. At field days in Australia, I would show a property's plant sequence by collecting the diversity of plants and then laying them out in an appropriate succession, which always opened up good discussion. If you are straight away wanting to also see the opposite plant sequence in a descending order, take another journey on where the globe is going if weeds are not valued for their reparative and regenerative role, please go to chapter eight and read – The Journey of Weeds. So, far in our voyage together, we have covered in the first three chapters as to why and how I formed my non-traditional viewpoints about weeds. Chapters four and five explained what plant succession was as an accumulative plant succession process and that it was bi-directional. The global reality is that there are more degrading landscapes and associated plant successions than improving ones. As I aligned my on farm pasture observations of nature with plant succession that took us through into this chapter of an example of a positive ascending plant sequence.

The following chapter was personally hard for me to write as changes in life will and do happen. You are going to journey with me to where I will introduce a new term: **positive feedback loops**, which describes a weed's functional ability to advance the plant succession in a regenerating landscape. The term helps to explain how plant succession is self-perpetuating and accumulative. It is as if these positive feedback loops drive the progressive plant succession. Later I will use the term amplified positive feedback loops, as plants called weeds use unique characteristics or traits that are amplified when they are eco-triggered. It is a weed's amplified positive feedback loops, especially in degraded landscapes, that enables them to have such superior production and duplication that gives use cause to destroy them.

CHAPTER 7

WHAT MAKES A WEED, A WEED?

"Attempts to control weeds without addressing the causes of the invasion are doomed because they treat symptoms rather than causes."

Hobbs and Humphries (1995)

CHAPTER 7

WHAT MAKES A WEED, A WEED?

My learning journey of what makes a weed a weed, came out of tragedy after the passing of our oldest son, Alex at age 22 with cancer. Having taken a year off consulting and spending 3 months with Alex in hospital mostly overnight our family was a wreck. Looking back, I had mental challenges and was in part suffering from being institutionalised in the hospital system for so long with my wife. A few days after Alex passed, I went for a walk where Alex and I used to go, but this time I was alone and the world looked very different. The trees looked different and I could not understand why. I had been looking at walls in a small room for so long, the world looked so big and very bright. I started to try and readjust to what the world looked like. I saw the world and landscapes differently. The trees were more like straight lines and their branches were like lines coming off them. I noticed different trees had different line formations and therefore different shapes. I could see the vase shape of shrubs and the ground hugging shape of bushes. I was relooking at trees in a different way. Then as the road got steeper, I could look up and see an elevated view and what caught my eye was that the grass on the side of the road was not even and seemed to have no defined boundary to it. It sort of wobbled and then grew for a while and then there were patches of no grass growth. Being right brain dominant, I am naturally inquisitive and my mind was trying to solve this natural puzzle. Then as I passed from shade into full sunlight, I felt the hot Queensland, Australian sun on my skin. In cool climate areas, plants do best in full sun and in shaded areas is where moss likes to grow. Where I live, half way up on the east coast of Australia, I was looking at the opposite, the grass was only growing in the shaded areas of the large trees (which we call '*bush*'). The environment of shade or direct sunlight was controlling the grasses establishment and productivity. Where there was shade the grass grew and in full sun there were just a few tap rooted 'weeds.' At the time my mind was still in a place being shocked by the big bright expanse of the great outdoors. I was struggling if I wanted to mentally pursue this line of thinking or not.

My mother's favourite earthy possession was an odd looking jug that had the words in light blue – "*Never say die, up man and try*!" Those words were in my subconscious and I decided to try and really understand what I was looking at. So, I stopped, stared and thought and then stared some more. I asked myself a question. "*What if each and every plant that I could see was the direct or indirect result of the environment that it was in*?" Not just one or two, but everything! Then I started to test my new theory and saw a tap rooted weed by itself in full sun doing very well, yet where there was a blanket of shaded grass there were no tap rooted weeds. Yes, acknowledge that there is such a thing as allopathy and some plant species produce allelochemicals as they grow, which inhibit other plants from growing. However, on the edges of the grass there were the weeds as the environment transitioned from shade to the full sun spectrum. As I looked around, I saw some lichen on the shaded side of a rock and that made sense. I saw tall grasses in a bit of a dip that collected water, that made sense too. I saw a weed in the concrete guttering where there was a crack as one concrete pour had stopped and then they did another section. The weed was doing really well as any rain came off the road and irrigated it, but how did the seed lodge there and get started in the

first place? These types of questions were racing through my mind and I was getting an understanding of the individual pictures of plants that I was looking at. I realised (apart from elevation) that the environmental conditions were the first dominating factor. I started to really identify that each plant was indicating something and all I had to do was to learn to read what it was showing me in greater detail. As I gained more clarity on matching individual plants with their preferred environmental conditions, I did a quick overview of the succession of plants that I was trying to interpret and again the weeds were filling in the gaps where grasses could not grow. I came to the bigger understanding that each healthy plant was there for a reason, it corresponded with the environmental conditions that suited it. The plant was in harmony with its desired environment and became naturalised. However, if the environment changed, so would the plants to again mirror and reflect that environment. Those events changed my thinking five years ago and as I better formulated my environment and plant matching ideas, I started to share these concepts by first delivering them at small '*Weed*' field days in a few states. As my confidence and validation of what I was communicating lifted, I then returned back to Key Note Speaking. This reminded me of when I used to be on the speaking circuit with Christina Jones, Maarten Stapper and, where I first met an interesting character, Peter Andrews, who greatly influenced my thinking and is the subject of the sequel to this book. I give credit to Peter Andrews OBE for delivering his understanding of the truth in a functional perspective of the Australian landscape. I would strongly recommend you read his first book – **Back from the Brink** and Chapter 16: Weeds are Allies, not Enemies.

As I did more public speaking about weeds, a sensitive way to give an alternative view of a weed was to describe it as an **indicator**

plant. My thinking deepened as I gave more talks and attempted to answer excellent questions. I am very appreciative of the landholders and agencies that over the years have pushed my boundaries, as I have also tried to push theirs. The harder the questions the more I had to think through the details and explanations, which for me created a rapid learning cycle towards the Judo Weed Management concept that requires a more in depth and different way of thinking. If I was at a field day and my engagement with everyone was going well, I would quickly transition my language from weeds being indicator plants to repair plants, but that had to be done in context to their localised plant succession. Before the start of a weed field day, I would spend a few hours collecting single plants and 'weed' species. During the day as a group, we would place them in some form of plant successional sequence. As I explained my interpretation of the sequence of plants, this always led to a lot of healthy discussions about localised conditions and specific plants, which I gained a lot of valuable local knowledge. I would tend to leave the '*problem*' weeds to the end of my overview of the local plant succession and ask the group what the role of the '*problem*' weed was and why was it there, just like Geoff used to ask me.

As my concept of weeds developed from indicator plants to repair plants, I had to establish how weeds repair the landscape and create a localised microclimate. I discovered that it is comparable to adding a new link to a chain, as after a disturbance weeds act like repair '*links*' in the plant sequence '*chain*' and their repair function is related to each weed species having an inherent feature of mirroring the limitation in their environment in degraded landscapes. To be very honest, trying to support the concept that weeds are repair plants has not been easy. What I observed and make sense in the

field, I could not find in the scientific literature, however, in the last few years it is improving (Tilley et al, 2022, MacLaren et al, 2020, Rogers, 2020). I personally got very down on myself over this difficulty in not finding usable scientific references and to try and cheer myself up I phoned Professor Stuart Hill. I complained to Stuart that I could find very little academic information about weeds as repair plants apart from his works (Hill, Ramsay, 1977). He said there was very little written down scientifically, only over stated generalisations. I shared that I was working on the assumption that there was a spontaneous bi directional succession of weeds. His response and indirect answer was to share a story about a Canadian called Fred Freeze who used summer fallow and sprayed his weeds. Fred used to say "*I am removing the weeds and God is growing them back*." This got Fred thinking so he started to let his weeds grow over summer and so they would not set seed, he would mow them and plough them in. Fred even asked individuals for weed seeds as they just used to feed them to the chickens. Over 15 years Fred slowly saw different weeds come in and then be taken over by another lot of dissimilar weeds. Fred saw a succession of weeds – heal his land. This greatly encouraged me. Peter Andrews (Andrews, 2006) shared a similar story of how he slashed weeds to create mulch and then he learnt the advantages of taking the mulch or hay to the higher points of the property. He suggested that the weeds are the best soil revitalisers. I knew that weeds were nature's choice, but it is far more complicated to explain. This concept of 'bad' weeds producing a 'good' regenerative result is counter-intuitive.

I have explained the transition from interpreting weeds as indicator plants to repair plants, and now the next transition is to explain weeds as producers of 'positive' feedback loops and the grouping of 'Weedy' Families. These weeds in each 'family' are related by the environmental conditions that eco-triggered them out of dormancy. I was trying to understand how weeds utilised their specific characteristics to compress time and space and advance plant successions in a way that other plants failed to do. What made a plant so productive that it gained our attention, so much that we labelled it a weed and then eliminated it? The simple answer was that weeds have the attributes that relatively quickly advance the plant succession to the point that humanity wants or needs to kill them. Are weeds just plants that are too good at their repair role?

As my study of weeds intensified, I increasingly stopped looking at individual weed species, not just as an indicator plant or even a repair plant, but at trying to identify the enhanced characteristics that gave weeds the relatively rapid ability to raise the successional order following a disturbance and coming out of inactively or dormancy.

What was causing this effect? I thought back to the many years I had spent with Peter Andrews including a 3,600 km return road trip between Mulloon Creek (Canberra, NSW) and south of Rockhampton (Queensland) when Peter had just come out of hospital. On that trip our learning group observed so much degraded drought affected country together with Peter chewing my ear (nagging me) and energetically waving a stick at me at Gerry Harvey's Baramul Stud. All these enlightening experiences planted the regenerative seed to write the sequel to this book; Reversing Climate Change Solutions – Water, Weeds and Why? Peter said that weeds were very important and, in his book; Back from the Brink (Andrews, 2006, pp. 96-98) he tells the story of how he used to talk about '*cause and effect*' and that Professor Willy Ripl corrected him to use the term 'feedback.' Was it possible that weeds were positively transforming 'their'

environment through what Professor Willy Ripl would describe as feedback loops? All natural systems have feedback loops as that is what, in part, makes them a system. What I was seeing was the effect of different positive feedback loops on enhancing the plant succession by weeds, but not fully understanding this natural process. I was transitioning from interpreting weeds as indicator plants to repair plants to the producers of 'positive' feedback loops. What I started to see was that the distinctive characteristic form and function of weeds could be interpreted as overall 'positive' feedback loops giving weeds an ability to raise or reset the plant succession in their bio-region and more specifically where each plant was actually growing.

By reading the weeds and interpreting their enhanced feedback loops in a landscape, I could also interpret and start to predict the historical (time) problems that caused the degrading environment that initiated the weeds in the first place. As I studied individual weed species, they expressed their own unique characteristics or traits in their form and function. These specific **Weed Functional Traits** (WFT) mirrored the degraded symptoms of their environment and localised conditions. Yes, weeds were still indicator plants, but they were now indicating the solution to fixing the environment that they were in. It was as if each weed had a role to play in fixing a degrading landscape and my task was to learn how to read each of their roles. After focusing on the environmental influences that were associated with weeds, I concentrated on the attributes of the weed that were contributing to making an improved environment where they were growing (macro environment) and their creation of a localised climate (microclimate). I tried to stop looking at weeds with a negative intent and cultural bias, but to observe what I actually saw. Again, back to Geoff's comments of – "*What do you* *see?*" I began to see the diversity of ways that weeds produced a 'positive' feedback loop in their physical forms (morphological) as they inherently mirrored the limitations of 'their' environment.

I started to specifically look at how weeds expressed and amplified their physical forms whilst displaying their surrounding environmental limitations. I was trying to understand the range of amplification of the positive feedback loops that finished in the death of the weed. The weed or generational weed's death brought to a completion its role in advancing the plant succession. Due to disturbance of the historical plant succession, weeds naturally are eco-triggered and what makes a weed a 'weed' is its ability to reinforce and amplify its positive feedback characteristics. (This is also one of the reasons why I used a magnifying glass on the front page.) I was starting to see weeds as producers of positive feedback loops which gave cause to their almost 'self-sacrificing' role as one weed superseded the other, just like in Stuart's story of Fred Freeze.

To use Arthur Sampsons words; "*The plants themselves, by adding humus to the soil through the decomposition of their tissues, and in this way changing the physical and chemical composition of the soil, prepare the way for a new and higher form of life, hence in a way work out their own destruction*" (Sampson, 1919, p. 3). Weeds seem to be speciality plants that can relatively quickly do that. Once these feedback connections were identified, I also began to see distinctive patterns with the most obvious being the two main root systems, tap roots (from a **dicots** or two first leafed plant) and fibrous roots (from a **monocot** or single first leafed plant). Again, please go dig up under your 'problem' weed and look at its root system. The second major pattern were the 'protective' weeds and the 'productive' weeds. Protective weeds are associated with

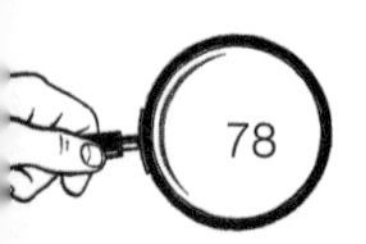

degrading landscapes and can be related to coming down the plant succession staircase. They included plants with armour, weapons and poisons. These plants are less productive as their energy goes into the adaptive traits. These plants would be on the lower side of your Desired Plant Landing (DPL). The weeds on the higher side of the DPL are often the 'Productive' plants and weeds that utilise excess nutrient and the common garden and pastural ones are nitrate weeds, as shown on the bottom of the 56 Naturalised Indicator Pasture Species in SE Australia for Acid Soils. The other 'Productive' weeds tend to be the environmental weeds that take advantage of or make niches (creating Positive Feedback Loops) with higher nutrient levels as this is the reason why they grow so quickly and are problematic. Most of the problems with Weeds of National Significance (**WONS**) is that they grow and spread relatively quickly!

An unintended educational breakthrough of explaining the feedback connections and distinctive weed species patterns occurred when I had been voluntarily teaching a group of my friend's children about agricultural and farming. In planning to see them again I wanted to share about the different weed groups that had different positive feedback loops. My challenge was to keep it interesting and relevant to young active minds. On the weekend of the educational morning, we covered the previous sessions and then went for a walk and saw a thistle with its thorn and called it the '**Armed Sword**' plant. We discussed what it looked like (form) and what was its role (function). We talked about how on heavier soils where their horses had over grazed, thistles had come up. We went into the bush (native wooded landscape) and saw some really rough native grass and I called it a '**Rusty Iron**' plant and that the soil was probably high in silica and that was why stock had not eaten it. It had good soil around its base and was protecting its small patch of ground. We saw where it was a bit wetter and there was a plant with three oval shaped leaves together and I said it was the **Bio-Lump** plant. This comment got a few weird looks and I explained that legumes fixed nitrogen and had lumps on their roots, which only a few of them had seen before. We saw where the clover was growing, the grasses were growing a bit better. As we walked from the lawn to the field into the bush, we finished in the main garden. We had covered a wide part of plant succession and I was enjoying linking visual names that represented the plants. When I got back home, I made a list of the plants we saw and started to create groups of related 'Weedy' Families by their form and function. This process took a few years and currently there are seventeen categories of related form and function 'Weedy' Families. The eighteenth category could be larger shrubs through to tree, but in general they are not described as weeds, however small 'tree' shaped plants could be described as woody weeds. I feel now could be a good time to introduce you to the crazy 'Weedy' Families. You have already met the '**Nitrate**' weed family in the bottom of the 56 Naturalised Indicator Pasture Species in SE Australia for Acid Soils. It would now be the appropriate timing to introduce you to all the 'Weedy' Families as they are a bit of a crazy group, being small and big, thin and wide, stocky and skinny and some even have poisons.

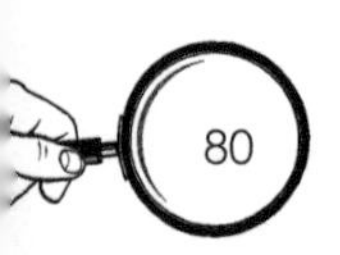

CHAPTER 8

THE JOURNEY OF WEEDS

"Each plant is an indicator.
This is an inevitable conclusion from the fact that each plant is the product of the conditions under which it grows, and is thereby a measure of these conditions. As a consequence, any response made by a plant furnishes a clue to the factors at work upon it".

Frederick Clements

(1920, p.28)

CHAPTER 8

THE JOURNEY OF WEEDS

You are about to take a very unusual bus journey and the story that you are about to read involves a **time line** from a beautiful green biologically diverse and complex **natural ecosystem** to a dry, soil less, plant less place called - *The* ***Desert***. You will be travelling on the **bus** of **human intervention** that over time slowly falls apart. The **driver** 'Mr Disturbance' who represents the creation of **human disturbance** has only one job and is task focussed. He decides to press on regardless of what is happening around him, until he is forced to run away. The **road** is the **ecosystem** that slowly degrades and the **named** gathering **places** reflect the changing **environments** where plant passengers board and leave. The **passengers** represent different **'Weedy" families** (biological communities) with their own special traits and amplified positive feedback loops that relate to and reflect their current environment. Eventually, the bus goes beyond - *The Place of No Return* where it is taken over by the last group of canopy making weeds (passengers) that set off to form a new pathway and create their own new road (novel ecosystem), which was not there in the past. These last passengers are what the new novel ecosystem makes. These self-organising environmental weeds are the ecosystem renovators making a new direction for themselves. Please enjoy an intentionally light hearted story about the very serious topic of the decline of a once healthy productive ecosystem and as it declines so will the fate of our civilisation.

The Journey of Weeds

The new bus driver, **Mr Disturbance** turned his new bus off the broad main road (natural ecosystem). Leaving the forested hills behind him, he went on a modern well-drained country road (disturbed ecosystem) to a gathering place called ***Ploughed Ground*** (environment). The first passenger group to be collected was the "**Pioneer"** family of weeds who were always the first to get on board. There was a lot of them and they very quickly filled the bus. They really looked as though they were in a hurry to get somewhere. As the driver Mr Disturbance progressed down the wide well-maintained all weather road (modified ecosystem), he dropped off part of the very diverse group of "**Pioneer**" family of weeds at a second gathering place called **Recovery** (environment) and picked up part of a new very large group called the "**Highrise"** family of weeds. They were tallish and were all-rounders and looked as though nothing would bother them. Then to one side was a brighter group called herbs and some carried a medical bag. As the driver Mr Disturbance went further down the road it looked **darker** and had **insects squashed** on it (simplified ecosystem). He stopped at the place called ***High Nitrogen*** (environment) and a lot of the "**Highrise"** family got off and a whole community with the **"Balloon," "Spider,"** and **"Umbrella"** families of weeds got on. There was the **"Balloon"** family who were very dark green, looking as though they had all eaten too much, whereas the long spreading **"Spider"** and big leaved **"Umbrella"** families looked as though they were all getting tangled up. As the driver Mr Disturbance went still further down the road it got **rough** and **bumpy** (dysfunctional ecosystem) and he arrived at a fourth gathering place called **Inferior Ground** (environment). Everyone got off at the stop, except for a few of the herbs, and a new family group called the **"Soil Tiller"** family of weeds came on board with their fine white roots. Heading off,

the road started to get very **windy and twisty** (depleted resource ecosystem) as the bus came to a fifth gathering place called ***Lesser Green*** (environment). Nearly everyone got off and the **"Bio Lump"** family quickly boarded the bus. They mainly had three rounded leaves with a few having wearing T-shirts with the capital letter N+ in a circle on them. Some were tall, some were short and wide, but some had runners.

The driver Mr Disturbance turned down a smaller **narrow** road (low biodiversity ecosystem). At a sixth gathering place called ***Over Grazing*** (environment), there was a rush to get off, just like no one wanted to stay around this place, except for a few stubborn hairy herbs on the back seat. There were two families that got on that looked as though they had been mining very hard ground. The **"Jack Hammer"** family were first, each carrying their own jack hammer with some having longer bits than the others. The very tall **"Drilling Rig"** family looked like they worked deep underground exploring new areas, because the soil was so poor. The bus moved off and the road's sealed surface started to deteriorate and break down (declining ecosystem services). The seventh gathering place was called ***Nothing on Top*** (environment). Most of the other two families got off, but one bumped her head on the way out, the short ones thought that was funny. The driver Mr Disturbance got out to check his bus and someone had written graffiti and flora art in the dust that was all over his bus. This next family had been waiting there for a while; they were in no rush to get on board as it looked as though they had lots and lots of luggage, it even had to go on the roof! The **"Gold Nugget"** family of weeds were very rich and had stored up their wealth.

The bus was slow to get going again and the driver Mr Disturbance had problems with the gearbox and clutch, and the back-bumper bar was falling off. He eventually turned off down a **thin side** road where the bitumen diminished and then disappeared to just gravel (overexploited ecosystem). He stopped at the eighth gathering place called ***Poverty Place*** (environment), before he realised the new passengers had boarded the bus and thrown the **"Gold Nugget"** family off. The **"Armed Sword"** family of weeds looked as though they were in fancy dress with thorns / spines / spikes (apparently, some also had a few pots of poison!). In this family no one got close to each other and it was one weed to a seat. Off the bus went and he took another turn on to a partly gravelled road (disintegrating ecosystem), then pot holes appeared (erosion) that got bigger (sheet erosion) and bigger (gully erosion) then a landslide had to be avoided. He dropped off the **"Armed sword"** family of weeds and luggage, and the bus was boarded by the **"Rusty Wire"** family of weeds. What could you say about this family? There was less of them, they were thin and very strong looking. The place he stopped at was called ***Not a Good Place to Bee*** (The locals spelt the last word – bee, as a joke as no one had seen a bee in decades). The declining gravel road worsened into a very cracked broken track (overharvested ecosystem). Most of the **"Rusty Wire"** family did not like the **"Armed Sword"** family, for they often left behind broken points in the seats, so most of them just stood upright with their long slender flowing hair in the breeze as the broken windows let the hot air through.

The driver Mr Disturbance had to go down to what appeared to become a slim **two wheeled dirt track** (denuded ecosystem). The bus was rattling and slowly moving as one of the back tyres was going flat. The bus finally came to a tenth gathering area called - **The Place of No Return** (environment) and the **"Rusty Wire"**

family slowly got off, but he had to wait a long time to board the "**Fortress"** family of weeds. There were only a few of this robust looking family with very thick legs. They slowly moved into the bus, as each had to carry a very large shield to defend themselves. In the bus they just sat on their shields and did not move. As the bus got going it seemed very empty, but was starting to overheat and had sand in the brakes as well as missing body panels. The driver could not turn around and slowly the track was getting covered with sand (collapsing ecosystem) until in the end it disappeared into an ocean of dry sand (ecosystem cessation) and his back wheels bogged. The driver Mr Disturbance came to a place called - ***The Desert*** (environment) and he can only blame himself for the directions that he took and for not turning back and finding a new route. Then in the distance there was a lot of very loud yelling and arguing. It happened, Mr Disturbance's worst fears!!!! A novel (landscape) weed family was arriving, which had travelled a very long distance to find the bus. The one weed family that he had deliberately avoided and left behind was now coming to the bus – The **"Transformer"** weed family that grow structured canopies. The driver Mr Disturbance switched off the engine, grabbed his back pack with his hat, compass, snacks, matches, water reserve (*plus some toilet paper*) and ran away from the bus. Then the huge group came over a sand dune and arrived with everyone pushing and shoving. They were a very varied family, many tall with branches, vines and canes. They spray painted on the side of the bus in dark green paint – POWER TO PLANTS. Then someone sprayed a lucky four-leaf clover on the back of the bus. A big fight broke out, and someone chanted – "Traitor! Traitor!" The clover was made into a black berry bush and everyone was happy.

One giant weed lifted the bogged side of the bus, another pulled the thorn out of the flat tyre, and yet another gummed it. When everyone was in the bus the head of the family got in the driver's seat, started the bus (which had cooled down) and turned it around very skilfully and carefully (some using their branches under the wheels) and started to drive as evening fell. Their big long **canopy** branches wrapped around the bus and held the panels together as they drove to form a new pathway where there were no official roads; they created their own new novel road (ecosystem renovation), which was not there in the past. Where do you think they were driving to or what do you think they were looking for?

P.S. You may be wondering what happened to the few "**Fortress**" family of weeds. They had heard the **"Transformer"** weed family coming, long before the driver Mr Disturbance, who was very, very worried about how his bus (on going human intervention) was falling apart and that his passengers (weed families) were getting rougher and fewer. So, the "**Fortress**" family of weeds jumped off the slow-moving bus, which had no doors, and each used their shield as a toboggan to escape. I forgot to mention that while on the bus they sat on their shields on the seats, so they did not sit on any spines or thorns - these were smart seasoned weeds!

This story would make an imaginative children's book with full illustrations. If you know of someone that would want to help children learn about naturally based solutions, please contact me. I have a separate Literacy and communication business as I have decoded the English language, so that you can sound out every sound, together with songs, poems and hopefully a new children's book.

There is, however, a serious side to this story for you have just read the transformation from a historical natural ecosystem to a "human-made" desert via unsustainable agricultural and farming practices. Each 'Weedy' Family that you met was trying to do their best as spontaneous plants to repair and restore environmental conditions, but as external conditions worsened, they literally could not hold their ground (soil). By learning to identify each 'Weedy' Family and why they are there you can better interpret your soil and ecosystem's health. Therefore, weeds are more than indicator plants as they are marking where your soil and environmental management is in relationship to the succession of plants in your garden, pasture, crop, lawn or other land uses. By learning to read the weed you too can take a trip into the succession of your own plants and go see what passengers you have?

Thank you for reading this little story, but remember: "**Plants show you where you are in the succession, but weeds tell you where you are going**." Some people have said that by reading the story a second time, they gained a better overview of what the story is getting across.

May I ask you a question?

Why is that weed growing in that place, what is causing it to grow there?

What can a weed's form and function tell you about my soil health and land management?

If I think back over time, has your weed community changed? (Why?)

What is going to happen when it goes from dry to wet or wet to dry, what weeds are you going to have?

If you were to be stopped at the place called High Nitrogen (environment) and meet the community of "Balloon," "Spider," and "Umbrella" families of weeds what would this be telling you?

What am I going to do when the next lot of weed passengers arrive?

CHAPTER 9

JUDO WEED MANAGEMENT

"Every plant is a measure of the conditions under which it grows. To this extent it is an index of soil and climate, and consequently an indicator of the behavior of other plants and of animals in the same spot."

Frederick Clements

(1920, p.3)

CHAPTER 9

JUDO WEED MANAGEMENT

Judo means Gentle Way and I believe that Judo Weed Management (JWM) expresses a gentle and eco-logical way to control and ultimately prevent weeds. I wanted to find and use a term that represented the soft art of weed control using **biomimicry** to achieve **Natural Based Solutions** (NBS). The art of Judo uses the principles of maximum efficiency with minimum effort, which is needed if our civilisation and associated political will is going to work smarter at weed control by having more of an eco-logical focus (ecocentric) and less of a top down technological focus (technocentric).

As a martial art, Judo includes manipulating the opponent's force against themselves rather than continuing force against them. Judo Weed Management is about manipulating the weed (opponent) against itself rather than spraying it with a herbicide and if the conditions have not changed the same weed species will repeatedly come back and attack you. In fact, if the weed is suited to the environmental conditions, it can build up herbicide resistance.

A (newly) disturbed environment eco-triggers weeds by creating niches and new opportunities for weeds to spontaneously grow. These (new) weeds match or mirror their preferred localised environmental conditions. Each successful weed has specialist adaptive functional traits, which I describe as **amplified positive feedback loops**, that enable the weed to outperform other plants and therefore give it its productive characteristic that makes it a weed. Each weed's specific amplified positive feedback loop enables the weed to contribute to the advancement of the self-perpetuating plant successional process. The greater the amplification of the feedback loops the more obvious they are. Big thorns are more obvious than small spines, which are less obvious than coarse hairs. Taller weeds are more obvious than shorter ones and green weeds are very obvious when other plants are dried out and dead. The form of the weeds represents the weed's function and by reading the weed's form you can interpret its function and role. The methodology of Judo Weed Management in part comes from comments that Stuart Hill once said that if you kill off the weed then you have to do its role. The role of the weed is to repair and then advance the self-perpetuating plant successional process.

The art of Judo management revolves around biomimicry and by using biomimicry management (**doing what the weed does**), which supplements the weed's original role. The functional role of the weed is displaced or substituted and the weed becomes dysfunctional, declines and dies. In general, the closer the biomimicry management can be achieved, the role of the amplified positive feedback loops, the quicker the dysfunction of the weed occurs. Therefore, the quicker you can work out how to biomimic the role of the weed, the greater the likelihood of turning the weed off.

I was once fully explaining the concept of biomimicry for weed control to a land manager and as I drew breath to give more detail, they said: "*So you just do what the weed is trying to do, but faster*" and that sums it up! An example is that if a weed is eco-triggered by the removal of (plant) ground cover, the Judo move is to restore it. If a flat weed is being eco-triggered by cutting your lawn too low,

then the Judo move is to raise your cutting height. If a weed is eco-triggered by an 'artificial' hard pan, then the Judo move is to rip just under the hard pan. If a weed is eco-triggered by compact clay soil, then an appropriate soil amendment is used as a Judo move. These management processes biomimic the advancement of the plant succession.

Judo has two main competitive objectives: to throw and immobilise an opponent. The second objective is to hold down a weed (opponent) to neutralise it. The holding down or preventing a weed is achieved by maintaining an environment and local conditions (position) that are not eco-triggering the weed out of its inactive or dormant state. These two objectives can be related to high soil nitrate levels where nitrate weeds are eco-triggered. Two Judo moves include either to add more plants (over planting or sod seeding / pasture cropping) to use the excess nitrate up or to add a small amount of sugar to the soil to feed the soil biology as they feed first. The addition of a small amount of sugar (Prober et al, 2005) can set back and prevent nitrate weeds.

Judo Weed Management (JWM) is not a stand-alone method of weed management it comes under the concept of Integrated Weed Management (IWM), which has expanded from Integrated Pest Management (IPM) using biological, chemical and mechanical methods to include cultural and now ecological methods (Gage, Schwartz-Lazaro, 2019). JWM is more aligned with cultural weed management, but its features include learning to read the specific features of a weed and going to the advanced level of placing them within the regional plant succession. JWM strategies aim to identify where your Desired Plant Landing (DPL) is <u>within the regional plant succession and identify which are the most problematic weeds</u>, especially if and when there is a major change in the regional plant successional order. Judo Weed Management's objective is to take an eco-logical approach to weed management as it aims to have a logical approach to ecology that acknowledges and works with the self-perpetuating, self-organising natural phenomena of plant succession. Judo Weed Management is a more sustainable and eco-logical group of strategies for managing weeds and plant communities. In order to utilise biomimicry management as a way of creating naturally based solutions, we have to match what the weed's eco-logical role with the reasons why it is there and what (were) are the local conditions that eco-triggered them. By doing this and gaining an understanding of each 'Weedy' Family's functional role and its amplified positive feedback loop characteristics, this gives you an insight to each 'Weedy' Family's individual descriptive name.

Now to make the concept of Judo Weed Management start to work by turning weeds into solutions. We have to match the 17 'Weedy' Family members with the initial three **Jones' Weed Functional Groups** (JWFG) being 1] Ground Cover, 2] Raising Soil Fertility and 3] Nutrient Regulation. The first being the purpose of covering the ground (plant focused), next to purpose to raise and accumulate soil fertility (soil focus) and finally regulate nutrients (nutrient focus).

Jones' Weed Functional Groups (JWFG)

1] **Ground Cover** (Lower order plant succession):
Coloniser, Pioneer, Gold Nugget, Fortress, Rusty Wire, Armed Sword, and Tent Bush, which later goes into the Ecosystem Renovator group.

2] **Raising Soil Fertility** (Lower to mid order plant succession): Jack Hammer, Drilling Rig, Highrise and Soil Tiller (Grasses)
3] **Nutrient Regulation** (Higher order plant succession): N Lump, Balloon, Spider, Umbrella, Rope & Cloak and Bowl of Soup.

The above descriptive weed names list does have an official title being the: Jones Form Weed Index (JFWI).

As I wrote The Journey of Weeds story, I wanted to bring together a number of themes being the environment, ecosystems, their positive feedback loop characteristics or traits that are represented by their visual form names that represent their function. I now want to initially focus on their visual form and match them with the degraded local conditions that often eco-triggered them in the first place. By doing this you will be able to read the 'Weed' Family and relate it to the causative effect being the (historical) degraded localised conditions. The opposite will also be true that if 'problematic' localised conditions are intentionally or unintentionally created, you will be able to more accurately match which related 'Weed' Family will be eco-triggered. I am going to show you how to match the weed (symptom) with the local condition (cause) or the other way around. Previously I divided the 'Weedy' Family into three Jones' Weed Functional Groups (JWFG). Firstly, the lower successional order plants that are being eco-triggered to functionally cover the ground and re-establish a plant community again. Secondly, the lower to mid successional order plants and the focus or amplification in their root systems to raise and accumulate soil fertility. Finally, the higher successional order plants regulate the nutrient and build on the ground cover plant groups and soil fertility plant groups.

"Established weed management systems are wanting to control, stop or even eliminate plant succession. However, in general this is not acknowledged or documented. Plant succession is the elephant in the weed management room."

Gwyn Jones

CHAPTER 10

WEEDS AND WHAT THEY TELL

CHAPTER 10

WEEDS AND WHAT THEY TELL

This is a large chapter as we start to match the form of the weed with its causative effect. The aim is to look at the list of the 17 'Weedy' Family members and identify which one most relates to your weed or weed community. The 17 'Weedy' Family members are in very large general groupings. You could have a problem with an individual thistle species or thistles in general. This would come under the Armed Sword weeds and include weeds with thorns / spines / spikes. Try and focus on the form and function of the weed and ask yourself "*Why are they there and what are they telling me about the local conditions*?" By learning to read the weed you can also relate it to previous events that would have eco-triggered them like historical droughts or floods etc and also the future direction of the plant succession. Weed management and control aims to transition the environment and local conditions to disadvantage the weed and in general to enhance the desirable plant that you want to grow or to raise the plant succession in a degraded landscape. Or put simply, work to put the weed out of business by doing more of its job using biomimicry management and or compress the time that is needed for the weed to complete its task. Manage for what you want being a healthy plant and associated healthy functional landscape.

In this next section we explore the big topics of weed function and the corresponding form of different weeds that I have given simple names to. The 17 'Weedy' Family members are formally known as the Jones Weed Form Index (JWFI), which I would like

you to consider as if you were looking at a plant succession. The plants' first role is to provide ground cover and then defend it. The second stage involves soil building and the emphasis is on the root systems. The third stage is dealing with excess nutrient in one area and these plants are moving it from the soil with the exception of excess soluble nutrients that are associated with the Bowl of Soup weeds.

Ground Covers

The first six weeds in the Jones Weed Form Index, cover and or defend the soil. Coloniser and Pioneer weeds are re-setting to the lower order plant succession having no or small root systems. Gold Nugget weeds are hoarders and keep their wealth (carb/o/hydrate) underground as it is safer. Fortress, Rusty Wire and Armed Sword weeds are physical 'Guardians' of the soil with modified leaves and stems to protect themselves and guard 'their' precious soil underneath them.

Soil Builders

Jack Hammer, Drilling Rig, Highrise and Soil Tiller are soil building plants. N Lump being legumes can build soil nitrogen.

Excess Nutrient Users

The next five plants are very different as they are fast growing and productive weeds. You have to think differently about them and focus on what is being supplied to give these plants their production abilities. Don't just war at these weeds, but look at their supply chains and how that could be better used or diverted. Balloon, Spider and Umbrella plants are nitrate weeds and the next three being Rope & Cloak and Bowl of Soup. Tent Bush weeds are included here as they are also ecosystem renovators or I describe all six as transformer weeds as they are making and transforming localised novel ecosystems.

JONES WEED FORM INDEX (JWFI)

Jones Weed Form Index	Weed Function*	Weed Role
(Cover the Ground)		
Coloniser weeds	Sponge (scab)	*Re-setters*
Pioneer weeds	Small and rugged	*Re-setters*
Gold Nugget weeds	Underground storage	*Hoarders*
(Defend the Ground)		
Fortress weeds	Thick wall slow growth	*Guardians*
Rusty Wire weeds	Very tough leaves / stalks	*Guardians*
Armed Sword weeds	Thorns / spines / spikes	*Guardians*
(Build the Soil)		
Jack Hammer weeds	Short with tap root	*Soil Builders*
Drilling Rig weeds	Tall with deep root	*Soil Builders*
Highrise weeds	Thin tall & mophead	*Soil Builders*
Soil Tiller weeds	Fine roots	*Soil Builders*
(Build Soil Nitrogen)		
N Lump weeds	Fixing nitrogen	*Soil N Builders*
(Use Excess Nitrogen)		
Balloon weeds	Broad leaf growth	*Nitrate Weeds*
Spider weeds	Long thick spreading	*Nitrate Weeds*
Umbrella weeds	Long hollow spreading	*Nitrate Weeds*
(Use Excess Nutrient)		
Rope & Cloak weeds	Transfer of bulk nutrients	Enviro Weeds
Bowl of Soup weeds	Filters soluble nutrients	Enviro Weeds
(Make a Home)		
Tent Bush weeds	Makes canopy eco-house	Enviro Weeds

* The weed's specific restorative function is reflected in the its amplified positive feedback loops. It is what makes that 'Weed' Family unique.

MATCHING WEED PURPOSE TO WEED FORM / FUNCTION

Ground Cover	**Weed Form**	**Weed Function**
Cover the Ground	Bare / Disturbed Soils	
	Coloniser weeds	Sponge (scab)
	Pioneer weeds	Small and rugged
	Excess Leaf Removal	
	Gold Nugget weeds	Underground storage
Defend the Soil	Poor Ground Cover Soils	
Protective Weeds	Fortress weeds	Thick wall slow growth
	Rusty Wire weeds	Very tough leaves / stalks
	Armed Sword weeds	Thorns / spines / spikes

Build Soil Fertility	**Weed Form**	**Weed Function**
Mineral Replacement	Compacted Soils	
	Jack Hammer weeds	Short with tap root
	Mineral Depleted Soils	
	Drilling Rig weeds	Tall with deep root
	Highrise weeds	Thin tall & mophead
Topsoil Builders	Poor Soil Structure	
	Soil Tiller weeds	Fine roots
	Nitrogen Depleted Soils	
	N Lump weeds	Fixing nitrogen

Excess Nutrients	**Weed Form**	**Weed Function**
	Excess Soil Nitrogen	
Nitrate Weeds	Balloon weeds	Broad leaf growth
	Spider weeds	Long thick spreading
	Umbrella weeds	Long hollow spreading
Ecosystem Renovators	Sinks of Excess Nutrient	
Transformer Weeds	Rope & Cloak weeds	Transfer of bulk nutrients
	Bowl of Soup weeds	Filters soluble nutrients
	Tent Bush weeds	Makes canopy eco-house

The previous diagram has brought together three themes being the role, form (name) and function of weeds in general. As there are globally thousands of different plant species called weeds, the diagram is an overview designed for informed decision making. The above two diagrams and the information contained in this publication have been formulated in good faith, the contents do not take into account all the factors which need to be considered before putting that information into practice. Accordingly, no person should rely on anything contained herein as a substitute for specific professional advice. I specifically mention this as weeds can quickly come and go or become naturalised. Weeds are part of a dynamic living ecosystem. Controlling and managing weeds is in reality trying to control, manipulate and manage plant succession. I have given each of the 'Weed' Family members a name in relationship to their physical appearance, that also relates to their function form. By focusing on the weed's name, I am also highlighting its amplified positive feedback loops that make it distinct. Read your weed's functional role, then enhance or amplify that role to make the weed dys-functional. By reading the weed, you can turn your weed into solutions. **The Jones' Weed Functional Groups** (JWFG) can be divided into 1) Ground Cover, 2) Raising Soil Fertility and 3) Nutrient Regulation, which can also include them accumulating specific elements. Plants' first role is providing ground cover and some weeds physically defend it. The next stage is soil building and the emphasis is on the root systems. **Remember 50% of the plant can be underground**. (Don't forget the shovel!). If ground cover is achieved and soil fertility raised the higher order plants can function and deal with excess and too much nutrient in one area. We will commence with the functional purpose of ground cover.

1 GROUND COVER

A lack of ground cover is a major eco-trigger for weeds as it is a un-natural outcome and represents the early stages of a primary succession. Ground cover protects the soil surface and acts as an insulation blanket. Ground cover can be living (plants) or dead material (dung, stones, standing summer feed etc). Ground cover in the form of dead organic matter is a vital food source for soil biology, which plays an important role in recycling nutrients and maintaining or rehabilitating biodiversity. If ground cover is substantially lost there is an increased risk of accelerated soil erosion occurring. In this section of the book, the main weeds are those that cover the ground and those that defend the soil. Ideally the soil has a cover of decaying organic matter between the mineral soil and the living plant. This layer is the roof to protect soil biology and to feed it.

1.1 Covers the Ground

A primary purpose of plants is to cover the ground and in so doing protect the soil and create organic matter, which when decomposing creates a recycling feedback loop. The disturbance or removal of ground cover eco-trigger weeds and in very highly degraded sites the first weeds are colonisers having no roots, flowers or seeds. Coloniser weeds literally have to restart the plant succession.

Bare / Disturbed Soils

"*Here comes the first wave of pioneers, help is at hand*!"

The plants for these soils are *"weed"* colonisers that create the needed cover. These conditions trigger the seed bank of opportunistic "pioneer" plants which come in a range of sizes, forms and shapes. On lower fertility landscapes they are often the first out of the ground and quickly take up available nutrients and then cover the soil with a "green" scab, just like if our skin is damaged.

These can initially be short cycle, quick succession (annual) plants. In general, if you see moss growing (has no roots) there is a major soil problem.

1.1.1 Coloniser Weeds (Lichen, Liverworts, Mosses)

Coloniser weeds amplify positive feedback loops involve the formation of a sponge or soil scab at the start of an ecological succession. The name 'Coloniser' means to establish and connect to the ground. Colonisers are the first and act like little sponges having no roots, or flowers or seeds. In colonisation, initially lichens grow (withstanding dryness) and then mosses that need more moisture. They are short and form biological soil crusts that initiate soil formation. Lichen, liverworts, mosses are **cryptograms**. Lichens are not a plant, but have a symbiotic relationship between fungi and algae or cyanobacteria. Technically the latter can be the first to colonise, being just below the soil's surface. In secondary succession (agriculture) mosses are associated with surface soil compaction and surface sealing, forming a crust or scab to cover the soil surface. Have you noticed where repeated herbicides have been used and that only moss will grow? The eco-logical control of colonisers includes raising the plant successional order by improving the localised environment or to replace their role by regenerating additional ground cover.

1.1.2 Pioneer Weeds (Winter Grass, Toad Rush etc)

Its positive feedback loops are very fast from roots to tops then rapidly to a lot of mature seed. It has a lot of shoots, but relatively little root. Its feedback loops are small scale and has limited amplification as they are small in size and limited shallow root development. These plants are rudimental and rugged with a basic lack of refinement. They are crude looking plants often with shorter thick stems with small leaves. The name 'Pioneer' represents who

they are being the first to establish being small herbs and forbs coming in all shapes and sizes. In cropping, if this type of plant is an initial problem, the focus can be on a pre-seed treatment for example beneficial soil biology, compost teas or additional nutrient coatings etc. This is to give the crop seed a greater advantage in competition. By increasing overall landscape fertility and hydration Pioneer weeds can become dys-functional.

Excess Leaf Removal

The excess removal of leaves from plants weakens the plant's root system as root carb/o/hyrate (Carbon / Oxygen / Hydrogen) reserves are used to grow new growth. As this new growth is removed there is an overall loss to the plants. If this happens repeatedly the root system weakens and can die, especially in dry conditions.

1.1.3 Gold Nugget Weeds (Bulbs, tubers, corms like Onion Weed, Oxalis, Garlic and rhizomes like Sorrel and Bent Grass etc).

This is a very diverse group and could be further split up, but their amplified positive feedback loops involve root or stem storage and sometimes it is hard to work out which is which. Underground root storage vessels or stems can survive in adverse environments and they specialise in surviving long seasonal periods of stress when their tops are continually being removed. The speciality name of "*Gold Nugget*" is due to their main energy (gold) reserves are underground like bulbs, tubers, corms, rhizomes as well as above ground stolons or thickened runners. These plants are often the result of overgrazing / mowing, when other plants die out, they are able to quickly recover after rain and dominate. The eco-logical approach to these weeds could include increasing plant competition by letting all plants grow and out shade them in the right high production seasonal conditions.

1.2 Defend The Soil

"Defenders of the soil" are protective weeds and as their name implies, they use their physical form to protect themselves and for the soil's benefit. They are true guardians of the soil and use thorns, spines, spikes, hairs etc. Plants on more degraded locations simply use their inedible mass and put a coarse horizontal matting on and in the soil's surface. These plants often accumulate silica and potassium and use their physical inedible mass to protect the soil as a last resort. Their enhanced resistance to decomposition slows decomposition and gives a longer lasting surface coverage. Plants as weeds will also defend themselves by utilising chemical warfare = toxins and poisons!

Poor Ground Cover Soils

The local conditions of poor ground cover often occur where stock congregate or where crops / pastures have been sprayed out. Very low fertility and arid sites can also have poor ground cover with low decomposition rates and are more vulnerable. Note that poor ground cover can literally be in between inedible "Fortress" or "Armed sword" like weeds.

1.2.1 Fortress Weeds (Serrated Tussock, Spiny Rush etc)

Its amplified positive feedback loops involve being thick walled and slow to grow. If you were an ant it would look like a fortress to you! The speciality name of 'Fortress' comes from its ability to defend the very ground that it is based on and it will be there a long time. These weeds have thick round or flat armour that is very tough. They create walls of thick coarse tubes and densely cover small areas of ground. The function of these fortress-like weeds is to 100% protect the ground and be unpalatable so stock will leave them alone. You can chop, kick and burn these plants, but with rain they just come

back as they are often very long lived. They can be rushes and tussocks and some very old coarse grasses can look like them as well. The eco-logical approach to these weeds again could include increasing plant competition and raise the plant succession, but their control can be a very slow process. Slashing and or very high stocking rates can set back these plants, but if done in the wrong conditions it can trigger a new generation of these weeds. There are no fast answers for these slower growing weeds.

1.2.2 Rusty Wire Weeds (Ratstail Grass, Spear Grass, Love Grass) These are lower order 'protective' grasses and that is their function. Their amplified positive feedback loops involve having very tough leaves and stalks. The speciality name of 'Rusty Wire' is because it acts and can feel literally like wire, is very slow to decompose which is part of its long-term function. These physically tough plants can often occur after over grazing and/or the on-going inappropriate use of fire. There can be linkages between these coarse grasses and the use of fire. Fire is often used to reduce their competition, but it can also have long term negative effects. For example, to release and solubilise elements, especially potassium (which can be quickly leached). This can trigger a plant's adaptive defence mechanisms to increase their surface "toughness" with greater waxes and lignin etc. Grasses can respond with increasing silica levels that are more resistant to decay and provide more fuel for the next fire. Normally high potassium (low calcium) or high lignin plants, especially grasses, leave very coarse longer term organic matter on the soil. These weeds are mainly induced by degrading land practices and are associated with 'drought' affected landscapes. More research needs to be done on rehydration of higher areas to increase water and nutrient recycling as a starting point for working with these symptomatic degraded landscape weeds.

The eco-logical approach to these weeds is to manage what you want and focus on raising the overall plant succession. A farmer said once that when I took my herbicide budget and added it to my fertiliser budget, I made progress. He transitioned from treating and retreating the symptom (weed) and focussed on the problem (lower landscape fertility and functionality).

1.2.3 Armed Sword Weeds (Thistles, burrs, cactus etc)
These are not only 'protective,' but armed plants and their amplified positive feedback loops involve having thorns, spines, spikes even (silica) hairs etc. These weeds are armed and dangerous and their speciality name of Armed Sword represents the armament of the plant. They are telling you to stand away and leave them alone! It is a true guardian of the soil, keeping animals away. These plants are often eco-triggered by the seasonal baring of the ground or after a drought on lower fertility areas. Sheep and their ability to '*lip and teeth*' down to the ground, bare the ground more than cattle who more '*lip and tongue*' to graze. There is a relationship between sheep, over grazing and Armed Sword weeds. The eco-logical approach to these protective weeds could include increasing ground cover as an overall management strategy, particularly during drought periods.

2. BUILD SOIL FERTILITY

Having considered the importance of ground cover, which means that there are plants in the landscape. They are growing, dying and decomposing creating a nutrient cycle. The next successional progression is to raise soil fertility and overall soil health and productivity, especially on degraded landscapes. This is an essential regenerative tool towards regenerative Agricultural practices. The

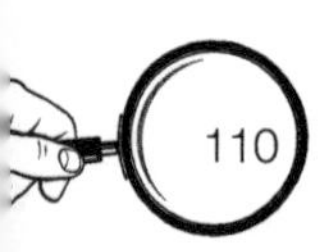

next group of soil fertility builders includes the plants that go deep in the soil and they can start off slowly as they focus on getting their roots down.

2.1 Mineral Replacement (Bio-indicator Weeds)
The use of weeds as Bio-indicators and mineral accumulates (McCaman, 1994) is well documented please see: Cocannouer (1950), Hill, Ramsay (1977), Carlesi, Bàrberi (2017), Maughan, Dominic (2021) and the classic book: **Weeds and What They Tell** (Pfeifer, 1975). Acres U.S.A. also has published a lot of related articles (Walters 1991). A feature of many mineral 'reaching' plants is their main central tap root. They have to go mining for new minerals, so they include the big 'Drilling Rig' weeds. This means that they are obvious as the landscape dries out and being the only green in the landscape they "*must be a weed*." These plants can act like living drill bits that help open up and drain hard pans. On their decomposition water is able to penetrate through the hardpan and the more desirable plants can then use these new entry points, enlarge them and start to break up thin (plough) hardpans. Their dead roots can also act like wicks allowing capillary water to move through the soil profile.

Compacted Soils
These local conditions include compacted soil, associated with clay subsoil on duplex soils or physically damaged soil with hard pans. These conditions can also include compaction on the soil surface. These plants use their jack hammer tap roots to drill down and find existing cracks and enlarge them. The weed herb dandelion has a dual role with a tap root and also accumulates calcium. Clay soils are traditionally low with calcium and additional calcium can "open" up a clay soil. The weed herb chicory is known for its deep hard pan

cracking and mineral up lifting. Compacted soils can be minerally rich but unavailable to plants roots, so often their leaves are soft and can be quickly recycled as an effective feedback loop.

2.1.1 Jack Hammer Weed (Flat Weed, Dandelion, Plantain)
Its amplified positive feedback loops involve a short thick with a tap root. The speciality name of "*Jack Hammer*" is that it breaks up compacted soil. The shorter version of drill rig with long flat leaves (handles) and a tap root (drill bit). Associated with compact soil, they can be clay breakers. The eco-logical approach to these weeds includes increasing the nutrient cycling to speed up the nutrient cycling that this weed is trying to achieve. These weed types build from the subsoil and re-mineralise the depleted topsoil, so corrective soil remineralisation can also act as a Judo move on them.

Mineral Depleted Soils
These local conditions can include shallow sandy soils (low CEC), highly leached soils (low calcium), over grazed or over cropped soils, repeated fodder conservation, highly eroded soils or simply just naturally low fertility soils. More conservative management is needed on landscapes with lower fertility as recovery rates can be a lot slower and during that time weeds can quickly dominate and create a more 'woody' weed cycle. These plants can be large and woody.

2.1.2 Drilling Rig Weeds (Red Dock, Chicory to 'Woody' weeds)
These weeds' amplified positive feedback loops involve being tall and having a long central root system that goes down deep. The speciality name of '*Drilling Rig*' reflects these tall weeds' ability to drill deep into the soil to find water and minerals. 'Woody' weeds and large trees are also in this group and their above ground size

increases their ability to go deeper into the soil's volume. These weeds can need a massive top in order to drill deep into the subsoil, so that new nutrients can be brought up to the surface and recycled. Woody weeds can be associated with potassium accumulation and recycling. They hold the potassium in their wood that returns to the soil or, in the case of fire, it is quickly released, but subject to leaching on the first heavy rains. A general comment is that in degraded landscapes the taller weeds often indicate bigger soil and biological problems. The eco-logical approach to these weeds is complex but includes increasing the nutrient cycling to speed up the nutrient cycling that this weed is trying to achieve. As landscapes degrade 'bigger' perennial weeds with deeper root systems can become the only resilient plants in that dying landscape.

2.1.3 Highrise Weeds (Fleabane, Farmers Friend)

One of the **most common general weed groups** and their amplified positive feedback loops are slender tall woody stems that on falling transfers fertility away from where it was first taken up. Its woody nature means that it will cover the ground for a while then slowly decompose. They are very much general all-rounders and have adaptive characteristics to not be too selective where they grow. I have also called these 'Mop Heads,' long sticks with fluffy tops that disburse their seed into the wind. The speciality name of the Highrise relates to their height and relative straightness. They cover bare spots, shade and protect the soil, plus bring up and recycle nutrients – an all-rounder. Some of these amazing plants can accumulate silica and this is reflected in their physical square stemmed structures. High silica means that they are hard to physically cut and stock's teeth can wear down. Remember silica is opposite to Calcium. The eco-logical approach to these weeds could include increasing plant competition while they are small and before they elongate. Slashing is often common as they 'stick' out being tall but have a small base area.

2.2 Topsoil Builders

This mainly involves the 'Grassy' Family that builds topsoil and improves soil health by improving soil structure and overall topsoil depth. They often simply fill in vacant spaces and germinate in small ecological niches. They are opportunists and improve the physical and biological components of the soil. They leave the soil in a better physical and biological condition. Please see: **Soil Health – The Journey** for a lot more information on building topsoil.

Poor Soil Structure

Poor soil structure can be highly related to a lack of soil biology and associated low levels of soil organic matter / soil carbon. Poor soil structure restricts root development and seed germination. It relates to a lack of soil aggregates. Developing soil aggregation in your soil, leads to the regeneration of your soil resource and in fact can make the soil younger. Soil Aggregation is about the joining or coagulation of small primary individual soil particles (i.e. sand, clay & silt) to form new structural units called, "secondary" particles, which scientists call "**peds**." These secondary particles are bigger in size than the smaller primary particles. Soil biology builds soil aggregates and so does fine root systems. Cultivational damage can cause poor soil structure together with excess nitrogen-based soil fertilisers.

2.2.1 Soil Tiller Weeds (Higher order 'soft' productive grasses)

These weeds' amplified positive feedback loops involve having relatively shallow fine roots that improve soil structure and soil aggregation. Their speciality name of '*Soil Tiller*' reflects their fine (surface) hair like roots that improve soil aggregation, structure and

soil health. When these plants are pulled out you will see how the soil clings to the roots. Most of these plants are soft grasses, but it can include the 'Highrise' weeds, especially with their surface roots. Grasses can be one of the hardest weeds to control as they are so competitive and can be climax plants in drier areas. The ecological approach to these grassy weeds can include the weakening of their root system by cutting or grazing them very short and as soon as green shoots come up and before the plant can start to rebuild its root reserve, the shoot / leaves are cut again. This process is repeated, weakening its root system. This is one method of weakening grasses in the interrow to limit their water and nutrient uptake at critical times. What I am describing is literally over grazing, but then Gold Nugget weeds could come in as a replacement. I am stopping the chapter here as there will be a change to the previous diagram as I need to introduce a new section on nutrient regulators. Previously we covered ground cover and building soil fertility, this subheading covers the next six plants, but not Tent Bush as it is an ecosystem renovator.

EXCESS = DEFICIENCY

"Do you have too much nutrient or a deficiency of living plants and biodiversity?"

Gwyn Jones

CHAPTER 11

NUTRIENT REGULATING WEEDS

CHAPTER 11

NUTRIENT REGULATING WEEDS

This is a separate chapter as I need to introduce a subsection of nutrient regulating weeds. It is the natural progression from plants covering the ground, to building soil fertility and then regulating the nutrients. Peter Andrews (2006) also mentions that plants manage the water and water-soluble nutrients cycles. There are two opposite functions in this weed (plant) group that regulate soil nitrogen*. I am using nitrogen* as a general term and deliberately not referring to ammonia, nitrite or nitrate or aminization, ammonification or nitrification to use some agronomy terms.

The nitrogen accumulators (legumes) and the nitrogen exploiters (nitrate loving) weeds. The first weed group are the legumes and they can regulate nitrogen levels as they can raise the nitrogen level of fixing atmospheric nitrogen when it is needed and especially after high rainfall (seasonal) leaching events. This is why you can have a good clover year and then afterwards, a good grass year that exploits the accumulated nitrogen. You can also see good healthy grass growing next to a healthy clover patch. The other nitrogen regulating plants are the nitrate loving weeds generally known as nitrate weeds. The third group are the enviro-mental weeds that grow rapidly due to excess nutrient. Enviro-mental weeds get our attention as they can grow, reproduce and start to quickly invade landscapes. That is also why I call them 'Transformer' weeds as they can transform landscapes and ecosystems. The last is the Tent Bush as it is an ecosystem renovator and not a nutrient regulator.

MATCHING WEED PURPOSE TO WEED FORM / FUNCTION

Ground Cover	**Weed Form**	**Weed Function**
	Coloniser weeds	Sponge (scab)
	Pioneer weeds	Small and rugged
	Gold Nugget weeds	Underground storage
(Protective Weeds)	Fortress weeds	Thick wall slow growth
	Rusty Wire weeds	Very tough leaves / stalks
	Armed Sword weeds	Thorns / spines / spikes
Build Soil Fertility	**Weed Form**	**Weed Function**
	Jack Hammer weeds	Short with tap root
	Drilling Rig weeds	Tall with deep root
	Highrise weeds	Thin tall & mophead
	Soil Tiller weeds	Fine roots

Nutrient Regulating Weeds		
(Nitrogen Builders)	Nitrogen Depleted Soils	
	N Lump weeds	Fixing nitrogen
Excess Nutrients	**Weed Form**	**Weed Function**
(Nitrogen Users)	Excess Soil Nitrogen	
Nitrate Weeds	Balloon weeds	Broad leaf growth
	Spider weeds	Long thick spreading
	Umbrella weeds	Long hollow spreading
Ecosystem Renovators	Sinks of Excess Nutrient	
Transformer Weeds	Rope & Cloak weeds	Transfer of bulk nutrients
	Bowl of Soup weeds	Filters/soluble nutrients

- - -

	Tent Bush* weeds	Makes canopy eco-house

Tent Bush* weeds Makes canopy eco-house

Nitrogen Depleted Soils

In general these local conditions include low organic matter soils, burnt soils, flooded (longer inundated), surface sealed soils, poor soil structure (heavy clay), biologically "dead" soils, very high organic matter soils (peat) and the classic of ploughing in / incorporated dried out low nitrogen crop residues. Common nitrogen depleted soils include over cropped sandy soil as they can hold very little available total nitrogen. Other soils include structurally damaged soil and anaerobic soils (impairing soil biology feedback loops) or excess carbon soil that have a very low carbon to nitrogen ratio. An ecological solution can include building soil protein (living organisms) levels and increase nutrient recycling.

2.3 N Lump Weeds (Clover / Medic, Broom, Mimosa, Wattles)

Nitrogen building plants are traditionally associated with the legume family of plants and for the purpose of this 'weedy' book they will be the main focus. They generally have (3) trifoliate leaves and their amplified positive feedback loops involve the speciality role of fix atmospheric nitrogen in symbiotic relationship with nitrogen fixing bacteria (rhizobia). The speciality name of '*N(itrogen) Lump*' relates to it having nodulation and nodules, which come in a range of sizes. The more common known legume plants, the ones we eat like beans, soybeans, peas, peanuts, and other pulses. Other legumes include White Clover, Vetches, Lupins and the 'King' being Lucerne that can raise soil nitrogen levels. On lower fertility soils; sub-clover, annual medics / trefoils and tropical legumes etc, can also regulate soil nitrogen levels. The diversity of these plants is large, one common feature is that being a legume they need vitamin B12, which comes from the synthesis of the mineral cobalt. B12 is essential for legumes and ruminant animals for cattle / sheep. This is one reason why B12 is added to Gwyn's Home Made Brew in

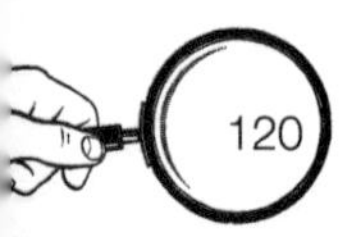

Soil Health - The Journey (Jones, 2008). Generally, legumes have a higher nutritional value and prefer higher soil fertility levels, especially phosphorous, calcium and the related element **boron**. They are associated with raising the order of plant succession, but are often preferentially grazed out with overgrazing.

As you develop up your own plant succession list see if you can identify (annual subterranean clover) legumes in the lower order as well as in the higher order perennial clovers. Remember that legumes are a major key stone plant as they can raise nitrogen levels with healthy nodulation that look like small bumps or knobs on the roots. When you dig up your legumes be careful not to knock of any nodules. Gently washing the roots can help and then cut open nodules and see if they are pink and healthy. In pastures clover can become dormant after wet seasons due to nitrogen leaching in the soil. Similarly, after a drought with high nitrate soil, you generally do not get a flush of legumes but nitrate weeds. **Practical tip**: Just because your landscape does not have legumes is not necessarily bad as they may not be needed.

3. EXCESS NUTRIENTS

Previously we covered the plants first role is to provide ground cover and then defend it. The second stage involves soil building and the emphasis is on the root systems. This third stage is dealing with excess nutrient in one area and these plants are moving it from the soil with the exception of excess soluble nutrients that are associated with the Bowl of Soup weeds. Excess nutrient in a landscape is not normally occurring in natural primary plant successions as it means that feedback loops have been disrupted and nutrients are 'leaking' out of the ecosystem. If there is a supply of available water soluble nutrients, there is normally a supply of living plants to utilise them, as nothing is wasted in nature. These groups of weeds are very opportunistic plants and are divided into 'Nitrate' weeds with 'luxury' feeding on high soil nitrate levels and Ecosystem Renovators that tap into and take up available nutrient. I think of 'Nitrate' weeds as the soil's 'Safety Valves' that are letting off excess nitrate and detoxifying the soil. The second group takes up excess nutrients and Bowl of Soup weeds actually trap and filter excess nutrient just like the human adaption of a reedbed sewage system. If there is excess nutrient it generally comes from somewhere, especially if it is in a soluble and transportable form. Excess fertiliser or polluting agents can cause excess nutrient. Degrading landscapes are prone to wind and water soil erosion with transported soil colloids being deposited into nutrient sinks.

Excess Soil Nitrogen

These local conditions include excess nitrogen fertiliser (when plants are too small to it take up), warm soils getting their first major rain event, green manuring, wattle clearing, high stock concentrations (stock camps / under shade trees / water points etc). Good healthy soils can switch to having excess soil nitrogen when the soil biology is temporarily set back by flooding (inundated) and or surface sealed soils. Higher fertility soils that have poor or damaged soil structure or shallow soils with high surface organic matter levels are good conditions to grow these weeds. Stock grazing on high nitrate plants, especially when they start to wilt, can cause **nitrate poisoning** with symptoms of glazed eyes, ears down, slow to lift and drop their front legs, heavier breathing, very loose manure, dirty tails and stock not wanting to graze nitrate weeds, leaving them to last.

The following 'Weedy' Families can in part be related to 'Nitrate' weeds at the bottom of the 56 Naturalised Indicator Pasture Species in SE Australia for Acid Soils. The two main exceptions that do not easily fall into the family weed grouping below is Barley Grass being part of the more productive grasses and Capeweed from South Africa. Both amplified positive feedback loops are their fine surface layered root systems and early maturity and seed set, they are 'Barbarian' weeds.

3.1.1 Balloon Weeds (Mellow, Pigweed, Nightshade & Amaranthus)
Amplified positive feedback loops are their fast growing green broad leaves that stretch and balloon out, hence their name. The speciality role of these "*Balloon*" weeds is to have fast growing broad leaves. They are often initially shorter plants that look like over inflated balloons with darker green high water content leaves. Their job is to literally pump excess nitrogen out of the soil, convert it into stable organic matter and then deposit it back onto the soil's surface. These weeds are often annuals and act like nitrate safety valves as they have a luxury uptake of nitrogen, which can kill stock. Avoid cutting and wilting these plants for hay as it concentrates the nitrate. The eco-logical approach to these weeds could include stopping nitrogen adding nutrient including composts, either adding more plants, over planting or sod seeding / pasture cropping, to use the excess nitrate up or to add a small amount of sugar to the soil to feed the soil biology.

3.1.2 Spider Weeds (Hogweed and Wireweed)
Its amplified positive feedback loops involve long thick spreading stems. Its speciality name is because it has a central stem and has radiating 'legs' similar to that of a spider. Spiders are relatively flat and are ground hugging being similar to this weed in its young

growth. Its functional role is to spread the nitrate from a centralised point and take it a distance away and spread it around on its death. '*Spider*' weeds have thick transport structures / runners, radiating out everywhere, to rapidly cover the soil. Their palatability often decreases as they get older. The eco-logical approach to these weeds could include stopping nitrogen adding nutrient including composts, either adding more plants, over planting or sod seeding / pasture cropping, to use the excess nitrate up or to add a small amount of sugar to the soil to feed the soil biology.

3.1.3 Umbrella Weeds (Melons)

Its amplified positive feedback loops involve prostrate vines with long hollow spreading stems that can transfer nutrient a long way, then they accumulate nutrients into their large fruiting bodies, creating new nutrient sinks. Its large light weight leaves shade the ground and give it, its speciality name of "*Umbrella*" leaf weeds. They often come up after droughts with high nitrogen soils. The eco-logical approach to these weeds could include increasing other plant competition to limit their development or sod seeding / pasture cropping, which cuts the runners and also uses up nitrate. I am not aware of the success of using a limited small amount of sugar on these weeds.

As nitrate weeds are often related with bacteria dominated soils and associated conditions. Dale Strickler in his book 'Restoring Your Soil,' documents the positive effect of Mycorrhizal Fungi (MF) inoculation on 'nitrate' and other weeds. In regard to weeds and soil biology I am limiting my discussion, but there our other authors that specialise more in this important, but still emerging topic.

3.2 Ecosystem Renovators

These are often called 'Environmental' weeds as they change their environment! I also call these amazing weeds, transformer weeds as they are making and transforming localised novel ecosystems. Often these weeds are renovating the landscape and quickly building biomass. Ecosystem renovating weeds can quickly create 'novel' or new ecosystems, however these changes to the local and regional environment can potentially come at the cost of potentially further disrupting or kill out 'native' plants and animals. Ecosystem renovating weeds are mainly big, long lived, non native (exotic) plants that naturally are invasive, due to their superior adaptations. A classic example is 'Wilding' conifers in New Zealand. The question must be asked as to why these plants are able to grow so fast, have nutrients for a long life and be able to use this to become invasive and literally ecosystem renovators. These plants are often not making something from nothing, but utilizing unexploited nutrients (sinks). They tap into nutrient supplies, growing fast large and have adaptive features to create a new micro and macro ecosystem. They change their environment and that is often why they are described as 'Environmental' weeds!

Sinks of Excess Nutrient

A nutrient sink is an area where (excess) nutrient accumulates and it is often in lower areas and they can act like a bathtub. They can also be considered as 'polluted' areas due to their artificially high levels of elements including potentially heavy metals. These local conditions can be highly diverse as it involves not only the volume / mass of nutrient, but more importantly the concentration (dilution) of nutrients. For example, a relatively small amount of organic matter and nutrient on (O and A horizon) a sandy shallow soil can have a similar effect of a high organic matter (thatch) on a deep loam soil. A common feature with these areas are their association with degraded landscapes and or water ways where nutrients have been deposited. These degraded sites often have had soil and solubilized nutrients moved from higher in the landscape and deposited in flatter areas. It does not take a lot of nutrient concentration to create what can be described as a nutrient sink. Once these Ecosystem Renovating plants access these nutrients, they can then go ahead to create a canopy that literally builds an improved ecosystem with more stable ground and air temperature, improved daily water cycles, accumulating organic matter, wildlife delivering dung and urine. They structurally build their own little city with an efficient carbon manufacturing and recycling factory.

3.2.1 Rope and Cloak Weeds (Rubber Vine, Cat's Claw Creeper) Their amplified positive feedback loops transfer bulk amounts of nutrients out of the soil and literally across the landscape. They are transferring excess nutrient in plant growth. The speciality name of Rope and Cloak weeds refers to their ability to climb (rope) or smother (cloak) other plants. These often thicker stemmed weeds rapidly climb or smother bushes / trees as their function is to try to increase the biomass above ground by recycling excess nutrients that have occurred. They are excess nutrient conversion units and that is why they are so fast growing. As a landscape dries out and literally starts dying, in its water flow lines there can be a relatively high level of nutrient compare to the higher areas. Ecosystem renovators can tap into this nutrient sink. Their role is to create feedback loops and reconnect the excess nutrient back into the landscape, due to their rapid nutrient recycling ability. It is nature's way of stopping the nutrient loss out of a "broken" ecosystem and spreading it above ground. These big weeds are problematic as they are so productive and an eco-logical approach to these aggressive

weeds could include the rehydration of dried out waterflow lines to dilute excess nutrients. The aim being to spread the nutrient loading to other more desirable plants and increase the overall competition within the plant community. Remember that they are the symptom, work out what is causing them.

3.2.2 Bowl of Soup Weeds (Reeds, Willows, Alligator Weed)
These are riparian and aquatic plants and in Australia millions of dollars have been spent on the Bowl of Soup weed called the Willow. One farmer said "*Now they have removed all my willows, I now know why they planted them in the first place,*" as he lost his bridge and could not access part of the property. The Bowl of Soup amplified positive feedback loops include filtering soluble nutrients. The speciality role of these wet weeds is to create an environment to slow/filter/use/recycle water soluble nutrients. These weeds store nutrients in themselves transforming where they live. Often, they create a mini dam or step and slow /retain / hold water forming a bowl, pond or leaky weir (Andrews, 2006) of initially "*soupy*" mix of nutrient rich sediment giving its name - Bowl of Soup. Nutrients sinks to the bottom, building a new bottom up ecosystem. There are a lot of aquatic ecosystem renovating weeds that grow in water flow lines, on the water's edge or in the water. You have to think differently about Bowl of Soup (wet) weeds and the eco-logical approach to these weeds can include focusing on what is being supplied to give these plants their production abilities. Don't just war at these weeds, but look at their supply chains and how that could be better used or diverted. This completes the excess nutrient 'Weedy' Family, which is part of the Ecosystem Renovators, but does not include Tent Bush as it is not a nutrient regulator.

3.2.3 Tent Bush Weeds (Blackberry, Lantana, 'Thorny' bushes etc) It is fitting to end the Jones Form Weed Index (JFWI) and its 17 'Weed' Family members with Tent Bush as it is one of the hardest to deal with due to their perennial longevity and protective function traits. An increase in Tent Bush plants can be a sign of increasing desertification and they could be used as bio-indicators of climate change. The Blackberry bush in the southern states of Australia are a similar form and function to Lantana in the northern regions. The amplified positive feedback loops are their stems and branches that make a canopy or tent that gives their speciality name of '*Tent Bush.*' A mature plant makes a canopy and eco-house including solar powered air conditioning by creating frameworks with their leaves being like bio-tiles as an outer covering, which protects and limits soil erosion. These are 'problem' weeds for graziers and ecosystem restoration projects and an Integrated Weed Management (IWM) program is needed as the successional transition on these long lived perennial bushes is very slow. The main problem is that they are **the symptom of a degrading landscape** and backward (regression) trajectory of a plant succession, which has eco-triggered them. Therefore, their role is to repair a degrading of the landscape and then advance it, but this is normally a slow process that commercially cannot be achieved within individual, community or government strategic timeframes. These are plants of disturbance and also include thorned bushes that domestic stock cannot consume. These weeds require a IWM approach and can include herbicide, mechanical and physical control, biological control and control by fire. In Lantana: Best Practices Manual and Decision Support Tool, 2009, p.67, Queensland Primary Industries and Fisheries makes a suggestion relating to plant succession. This is the use of scrambling or twinning legumes in thickets and then stock force their way into access the legumes, but Lantana can be

toxic to stock cause poisoning and photosensation, so grazing is not recommended. There are no simple eco-logical answers for these Tent Bush weeds and if the landscape is dying as these weeds can be just before "*the last three steps are MISSING*." There is a need for rehydration of the landscape and region to increase landscape function for Tent Bush to be reduced and its role to diminish. I will be taking this concept further in my next book: Reversing Climate Change Solutions: Water, Weeds and Why?

This small plant that you did not know.

One day a plant was starting to grow.
In a garden, starting out very low.
Then its flowers felt the warn sunlight.
Its flowers were oh, so bright.

As it grew with seed on high.
It matured and began to die.
This small plant helped many to grow.
This small plant that you did not know.

Was this a weed or just a plant?
Its seed fed many a hungry ant!
A lacewing ate many a pest.
A ladybird used it for a rest.

The bee enjoyed its nectar too.
Weed or plant, depends on your point of view.

Gwyn Jones

02 03 2019

CHAPTER 12

REGENERATIVE AGRICULTURE AND WEEDS

CHAPTER 12

REGENERATIVE AGRICULTURE AND WEEDS

Regenerative Agriculture is a worldwide movement, which gives new opportunities for agriculturalists and landholders to integrate new ways to regenerate their soils, farm lands, landscapes and entire (agro) ecosystems. It is merging new ideas about weed management and is creating new themes, new alliances with grazing and cropping methods including Savory's holistic grazing methods, Seis's 'Pasture Cropping' and especially American modified cropping methods including multi-species planting, cover cropping and crimp rolling. Organic Regenerative Agriculture is embracing a broad view of weeds that are part of their multi-species cover crops, part of grazing biodiversity and in many ways part of the regenerative process.

Currently, in Australia there is also a strong political motivation behind Regenerative Agriculture and its alignment not only to soil (health), but with soil carbon sequestration and its potential commercial returns. Regenerative Agriculture sends a positive signal to farmers and landholders of the requirement to actually regenerate agriculture as we know it today. The very term Regenerative Agriculture is an acknowledgement that agriculture needs to be regenerated from its degrading state. It is acknowledging that something is wrong and there is a need for (urgent) change towards more positive and sustainable outputs, including benefits for the environment.

Regenerative Agriculture is a way of thinking and can involve the formation of a different mindset or paradigm shift. It is not what

could be described as modern '*conventional*' agriculture, which is currently accepting and utilising '*chemical*' farming practices that originated after World War I (killer gases) and World War II transferring armament production into (nitrogen) fertilisers. After WWII the modernisation of farming practices was a move away from traditional and generational farming methods that were closely aligned with seasonal events and low resource inputs. Many traditional farmers and consumers believed that the increasing use of 'artificial' fertilisers, herbicides, fungicides and insecticides (including DDT) contributed to a decline in food quality. In response the Howard / Balfour Organic movement was established and developed into the British Soil Association, which commercially moved towards what is now known as 'Organic' certification to ensure some form of guidelines of production methods to satisfy consumers' potential concerns. Similarly, the 1924 Austrian born, German philosopher Rudolf Steiner founded 'Bio-Dynamics' in response to declining soil, plant, animal health and devitalised food. The Biodynamic movement has also moved into certification with an emphasis on the 'life force' of produce and not just it being 'chemical' free. Organic Regenerative agriculture is innovative and also aligns concepts with other Australian innovators being Mollison and Holmgren's – Permaculture, Yeomans' - Keyline and Andrews' – Natural Sequence Farming (NSF).

The topic of weeds has historically been associated with frustration, anger, annoyance and even warfare for many landholders, community members, government agencies and policy makers. Within the agricultural community there is ongoing debate about how weeds should be managed / treated / controlled / eradicated / exterminated and, in particular, if the repeated use of herbicides is part of a modern regenerative agricultural approach or not.

This topic defines the differences between Organic / Biodynamic systems, which do not allow the use of herbicides. However, the 'open' acceptance of different management practices is both Regenerative Agriculture's strength (popularity) and weakness (division).

Why Regenerative Agriculture manages weeds differently is that they acknowledge both the importance of environmentally responsible and regenerative land management as part of their written weed management plan, which can be an essential proactive tool in validating landholders (intended) compliance to external demands. In making this plan give your reasons, why and how you intend to monitor, manage, control or eradicate weeds. It is important to have evidence-based decision-making that underpins your approach to weeds and making your plan. A priority of weed management should be a risk-based approach including feasibility, likelihood of success and impact. Risk-based prevention and early intervention being considered to be the most cost-effective approach for managing weeds (Invasive Plants and Animals Committee, 2017).

We have covered a lot together in this book and the 17 'Weedy' Families is a lot to take in. If you are struggling to get a handle on what I am trying to communicate, try doing the following exercise. In the late afternoon find a patch of weeds and sit on the ground, so that you are getting into their zone with the sun to your back and softer light on the weeds. Remember to leave your phone behind and experience the moment. Get yourself comfortable and plan to stay there for a while. You may even want to get down at ground level. Start by just staring, observing and studying one weed for about 10 minutes. The more you look, the more you will actually see. Look at the angle of the leaves from the (main) stem, the leaf formation, the change of shape from younger to older leaves, the different shades of colour and look at its leaf debris on the ground.

Why you have to stay there a while is that your approach would have scared away any insects and you will have to wait for their return. Do they return to the weed or to your preferred plants, if so, why? Then just stare, observe, study another weed species, again for about 10 minutes. Then ask yourself what is the difference between the two weeds? Then look at both weeds and compare them to your preferred plants. What is the main difference between the form and function between the plants that you do and don't want? Do you know what roots system is under each of the weeds? If not, find out and wash the soil out of their roots in a large bucket.

Learn to become a weed detective, finding the clues that you need, so you can turn weeds into a solution. If you transition to think of all plants as environmental 'indicators' then plants are of value and especially weeds as they are telling you the future direction of where the plant succession is going. Lastly, please remember that weeds were once native plants that were given the opportunity to grow due to human intervention and disrupting the historical plant succession. As you walk away from the area that you observed and have a look around and see what other plants are there, think that all the plants that you can see are positioned somewhere in the plant successional order.

How would you feel about not only making up your own plant succession list, but also sharing it with some friends or go to your friend's garden, landscape or property and help them make one up. You do not have to use scientific names, just the names that you locally use (or even make them up) as it is for your benefit. Into the future I would like to see local **Landcare Groups take up the challenge of developing and sharing a regional plant succession list** as it is one way to get everyone on the same ecological page.

May I ask you a question?

What is the reason the weed is dominating in that area?

Is there a pattern that the weeds follow?

Why is this specific location eco-triggering those specific weeds?

Your answer to this is one of the keys to unlock the solutions to your weed problems.

Do the plants that you do want have a different root system to the ones you don't want?

If so, what can you learn from this and is this part of your solution?

Are the weeds more defensive or more productive?

What can you learn from this and is this part of your solution?

Are you dealing with season (annual) or naturalised (perennial) weeds that are at 'home' in this environment?

Now look again at the weed and think of it as an 'indicator' plant.

READ THE WEED.

Looking at the weed with an expectation that it has the answer to its own demise.

Ask yourself what is the message that the weed is signalling to you about this exact location?

What is this specific weed showing, what is its specific amplified positive feedback loop(s) or what does it do best?

What is the function of the weed, what could its role indicate to you?

What Judo moves could you use against this weed using biomimicry management to replace its role?

Do you have any high fertility zones, consider going and finding some '*nitrate*' weeds?

In concluding, the topic of weeds is a controversial one as one person may support the growth of a plant and another will call the same plant a weed. I describe a weed as a plant that has a perceived negative cultural value. In short, it is not the plant that is the problem. The plant is a neutral part of the discussion that is why weeds make excellent indicator plants. The so-called weed is a product of its environment or the local conditions that it thrives on. Weeds are like passengers taking a ride (MacDougall, Turkingto, 2005). It is true that some "new" unwanted plants have invaded areas utilising their superior adaptive positive feedback loops; they are the 'Symptom' not the 'Problem.' I would suggest that there needs to be a rethinking of the old model of focusing on the "weed" as the problem or for that matter even the soil seed bank. If the existing weed management model is not delivering the expected goals or actually allowing weeds to increase in number, is it not timely to question the need for a new and more relevant biomimicry relatable model that suggests that weeds are the 'Symptom' and not the 'Problem'? In fact, weeds can be strategically used as indicator plants of what the actual problem is. **The real problem is the ongoing and accumulative decline in the Australian landscape**.

Whilst running a rural field day on weeds, I asked an initial open question, "*What do you want to get out of today*?" A farmer replied straightaway, "*How to kill bloody weeds*?" I wrote that on the white board. Four hours later, he was talking about it was good to see how weeds can indicate things like over stocking. That was a big change, but it is only 6 inches, between ear and ear. One of the greatest changes you can make is to encourage landholders and their affiliates on side with **read the plants first, and not the herbicide labels**. There is such a thing as herbicide resistance and also landholder resistance. In the future it would be good to avoid both with a new innovative way by learning to work with nature and interpret nature's adaptive repair processes which includes what many call – weeds. Weeds are just plants that we as a community have not yet successfully tried to understand.

> "Prevention is always better than cure, as it is far more cost effective to prevent weed problems than to control established weeds."
>
> NSW Local Land Services South East, 2016, p.3., Preparing a Whole of Property Weed Management Plan, Local Land Services.

In completion, as a consultant my clients want to '*buy my eyes*.' They want to see what I am seeing. Having finished writing this original 'weedy' book, what I am bringing to the educational table is a visual way for you to see weeds and what a novel ecosystem can look like with weeds as part of its ecological characteristic features and biodiversity.

Thank you for allowing me to share some time with you as we have journeyed to explore a degrading novel ecosystem and met it's sign post weeds. I hope that I have empowered you to be justified to make the time and go and read your landscape and local environment in a new way, so that you too can now turn weeds into solutions by reading the form, function and characteristic of each weed. Lastly, when talking to Lady Eve Balfour in her home, she suggested that I read the book: Sand Country Almanac by Ralph Waldo Emerson. I will let him have the last word: "*What is a weed? A plant whose virtues have not yet been discovered…*"

ABOUT THE AUTHOR

GWYN JONES

Gwyn Jones, is an international author and educator who uses his experience in running a successful agricultural consulting business for the last 25 years, to show us in simple language how to take a new eco-logical approach and turn weeds into solutions! He is a pioneer in the formal education of organic farming and Natural Sequence Farming.

In 1999 Gwyn and Liz Clay developed and wrote Australia's first 16 day commercial organic conversion course. The overcoming of weeds was the second largest limitation to organic farm conversion. To solve this problem, Gwyn started his journey to learn how to turn weeds into solutions. He has gone on to predict bi-directional weed succession by interpreting the form and function of weeds as a reflection of their local environment. As a key note speaker and consultant, Gwyn encourages others to learn the skills to read their weeds and how to bring on the weeds' own demise.

Gwyn is a multi-talented author and has a literacy and communication consulting business which after 9 years of research broke the English language code. He is providing new opportunities to overcoming reading difficulties and pronunciation challenges through the new Jones Phonetic Alphabet.

In his spare time, Gwyn enjoys collecting old history books and rocks, as well as writing poetry. He is a keen traveller, having visited the USA, England, Jersey, Spain, Andorra, India, New Zealand, Kenya, Uganda, Vanuatu and New Caledonia. He lives on the Gold Coast with his wife Muriel. They have been married for 35 years, and have had three sons.

Gwyn's way of innovative thinking is different to most as he has the ability to stand back and see the big picture, read landscapes and read the weeds that are indicating the future direction of the plant succession. In short, Gwyn Jones is a truly unique problem solver.

FUTURE RESOURCES

Interested in Developing a Healthy Farming System?
Good regenerative land managers do not use chemicals because they have removed the need to use them. In effect, they have learnt how to solve their problems and remove the symptoms of a sick farming system, i.e. weeds, disease, high culling rate, low fertility, poor seed germination etc. These series of *five two day workshops aims to assist you in developing profitable healthy farming systems.

Presenter Mr Gwyn Jones
GradDip Sust Ag UNE, MRur Mgmt Studs Syd.

Managing your Soil Resource (2 days)

- How to plan the development of a healthy farming system.
- Identify the ABC of your soil, how it is made up and how to best manage it.
- Balance soil fertility, manipulation of soil pH for optimum plant growth.
- Introduction to the Albrecht soil balance method and its potential limitations and benefits.

Managing Soil Q.A. to achieve Sustainable Agronomy (2 days)

- Development of a Regional soil audit using comprehensive Soil Systems Analysis.
- **Four independent soil audits** and results are included, which are also added together to form the regional soil audit.
- Optimising corrective fertiliser treatments and changing nutrient availability.
- Relating plant and animal health to soil audits interpretation and QA.
- Developing a management plan for a quality assurance program of your soil.

(First four days are one course with three weeks between to allow for soil tests to return, which make up are regional soil audit as a bonus.)

SoilCARE: Managing soil for Australia's Farming Future (2 days)

- Introduction to *SoilCARE: Conservation, Aggregation, Regeneration & Energisation. Learn how plants grow their own topsoil.
- Understand the principles of regenerating your soils' health and productivity. How to make soil a renewable resource.
- How to use triggering mechanisms for "Building Topsoil."
- Explore ways of increasing soil health and vitality without adding fertilisers.

Managing Healthy Nutrition on your Farm (2 days)

- Introduction to the 5 and 10 Principles of AgriNutrition.
- The interdependence between Soil, Plant and Animal Nutrition.
- Problem solving nutritional limitations through AgDiagnostics.
- Healthy nutrition to improve and correct pests, weeds and diseases.

Managing Resilient Production Systems (2 days)

- Management principles of Regenerative Farming systems.
- Systems Thinking to developing Resilient Agriculture.
- The Four Foundation Production Systems (B-D, Org, Bio & PC).
- Three steps to sustainable production through transitional pathway program.

***SoilCARE** can be taken as two separate days, but the remaining courses have to be taken in their sequential order. Consider doing the stand alone SoilCARE two day course before committing to starting the larger programs. Courses are delivered in regional locations where there is sufficient interest.

NEW COURSE

Turning Weeds into Solutions

Presenter Mr Gwyn Jones

GradDip Sust Ag UNE, MRur Mgmt Studs Syd.

Director of Integrated Agri-Culture P/L

Ph 042 776 1191.

Email info@healthyag.com

You'll learn:

- Why waste money on herbicides?
- How to create your own plant succession list.
- Learning where your Desired Plant Landing is and why.
- Identifying your weeds main amplified + feedback loop.
- Why you have the weed in the first place.
- When to make the decision to pull the trigger.
- The secrets of learning to turn your weeds off!

This course can be delivered in regional locations where there is sufficient interest.

If you want a free digital copy of Soil Health -
The Journey (ed. Jones 2008), please email info@healthyag.com

OTHER READINGS

Andrews, P 2008, Beyond the Brink, ABC Books, Sydney.
Blair, K 2014, *The Wild Wisdom of Weeds*, Chelsea Green Books.
Cliff, A 2017, *The Value of Weeds*, The Crowood Press, Wiltshire.
Collins, P 2017, *The Wonderous World of Weeds*, New Holland Publishers, Chatswood.
French, J 2006, *Organic Control of Common Weeds*, Aird Books P/L, Australia.
Marshall, T 2010, *Weed, Harper Collins Publishers*, Sydney.
Millar, J 1995, *Pasture Doctor: A Guide to Diagnosing Problems in Pastures*, Inkata Press, Melbourne.
Miller, L & Nicholson, C 2020, *Visual Indicators of Soil Condition: Online Edition*, Meat & Livestock Australia.
Wall, K 2019, *Working with Weeds*, Kate Wall, Brisbane.
Smith, W 1983, *Wonders in Weeds*, The Camelot Press, Southampton.

Please do not criticise 'old' references being Goethe (1790), Darwin (1883), Warming (1909), Clements (1920) and Sampson (1919) until you have also read them as they often contain the 'whole' before reductionist science experimented with these new ideas and started to argue amongst themselves. "*Don't forget to dig into history*."

> **"Those who cannot remember the past are condemned to repeat it."** George Santayana

Please go to: https://soilandhealth.org to search for classic books.

REFERENCES

Andrews, P 2006, *Back from the Brink*, ABC Books, Sydney.

Invasive Plants and Animals Committee 2017, *Australian Weeds Strategy 2017–2027*, Commonwealth of Australia, Canberra.

Carlesi, S & Bàrberi, P 2017, *Weeds as Soil Bioindicators: How to sample and use data*, FertilCrop Technical Note.

Clements, FE 1920, *Plant Indicators: The relation of plant communities to process and practice*, Carnegie Institution of Washington Publication 290. Washington, DC.

Cocannouer, JA 1950, *Weeds: Guardians of the Soil*, The Devin-Adair. {On line version available}

Darwin, C 1883, *Vegetable Mould and Earth-Worms*, John Murray, London.

de Faur, R 1966, *Sheep Farming for Profit*, Reed, Sydney.

Gage, KL & Schwartz-Lazaro, LM 2019, *Shifting the Paradigm*, Agriculture, 9(8):179.

Goethe, JW 1790, *Versuch die Metamorphose der Pflanzen zu erklären*, Gotha. (English Version).

Hill, S & Ramsay, J 1977, *Weeds as Indicators of Soil Conditions*, EAP Publication – 67, McGill University, QC.

Hobbs, RJ & Humphries, SE 1995, *An integrated approach to the ecology and management of plant invasions*, Conservation biology, 9(4), pp. 761-770.

Jones, G 1999, *SOILCARE: Conservation, Aggregation, Regeneration, Energisation*, Integrated Agri-culture P/L, Mudgeeraba.

Jones, G (ed.) 2008, *Soil Health - The Journey*, Integrated Agriculture P/L, Mudgeeraba.

King, FC 1951, *The Weed Problem, A New Approach*, Faber and Faber Ltd, London.

MacDougall, AS & Turkington, R 2005, *Are invasive species the drivers or passengers of change in degraded ecosystems*? Ecological Society of America, vol. 86, Issue 1, pp. 42-55.

MacLaren, C Storkey, J Menegat, A Metcalfe, H & Dehnen-Schmutz, K 2020, *An Ecological Future for Weed Science to Sustain Crop Production and the Environment. A review*, Agronomy for Sustainable Development, 40, 1-29.

Maughan, C & Dominic, A 2021, *Weeds as Bioindicators: A Farmer's Field Guide Species Guide*, Pearl Moss Press.

McCaman, JL 1994, *Weeds and Why They Grow*, Sand Lake, MI.

Mitchell, E 1946, *Soil and Civilisation*, Angus and Robertson, Sydney.

Newman Turner, F 1951, *Fertility Farming*, Faber & Faber, London.

Northbourne, Lord 1940, *Look to the Land*, London, Dent.

Northern Territory Government 2021, *Northern Territory Weed Management Handbook*, Department of Environment, Parks and Water Security, Palmerston.

Pfeifer, EE 1975, *Weeds and What They Tell*, Rodale Press, Emmaus.

Podolinsky, A 1985, *Bio-dynamic Agriculture Introductory Lectures,* vol 1, Gavemer Foundation Publishing.

Prober, SM, Thiele, KR & Lunt, ID 2005, *Add Sugar and Kangaroo Grass and Burn in Spring - A Recipe for Success in Woodland Understorey Restoration*? Virtual Herbarium, Charles Sturt University.

Rogers, G 2020, *Desert Weeds: Personal Narrative on Botanical First Responders*, Springer, New York.

Sampson, AW 1919, *Plant Succession in Relation to Range Management*, U.S.D.A, Bulletin No. 791.c.

Scott, JK 2000, *Weed Invasion, Distribution and Succession*. In BM, Sindel (Ed.), Australian Weed Management Systems, pp. 19-38, RG & FJ Richardson.

Strickler, D 2021, *Restoring Your Soil*, Storey Publishing, MA.

Tilley, D Hulet, A Bushman, S Goebel, C Karl, J Love, S & Wolf, M 2022, *When a Weed is not a Weed: Succession Management using Early Seral Natives for Intermountain Rangeland Restoration*, Rangelands. 44. 10.1016/j.rala.2022.05.001.

Voisin, A 1961, *Grass Productivity*, Crosby Lockwood & Son, London.

Walters, C 1991, *Weeds: Control without Poisons*, Acres, USA, Kansas.

Warming, E 1909, *Oecology of Plants: An introduction to the study of plant communities*, Clarendon Press, Oxford, UK.

Wilson, A Jones, G Paynter, G Edser, G Norris, D & Kravcik, M 2023, *Hydrology, carbon and contours - the future of farming*, SCIREA Journal of Agriculture, Vol 8, Issue 2.

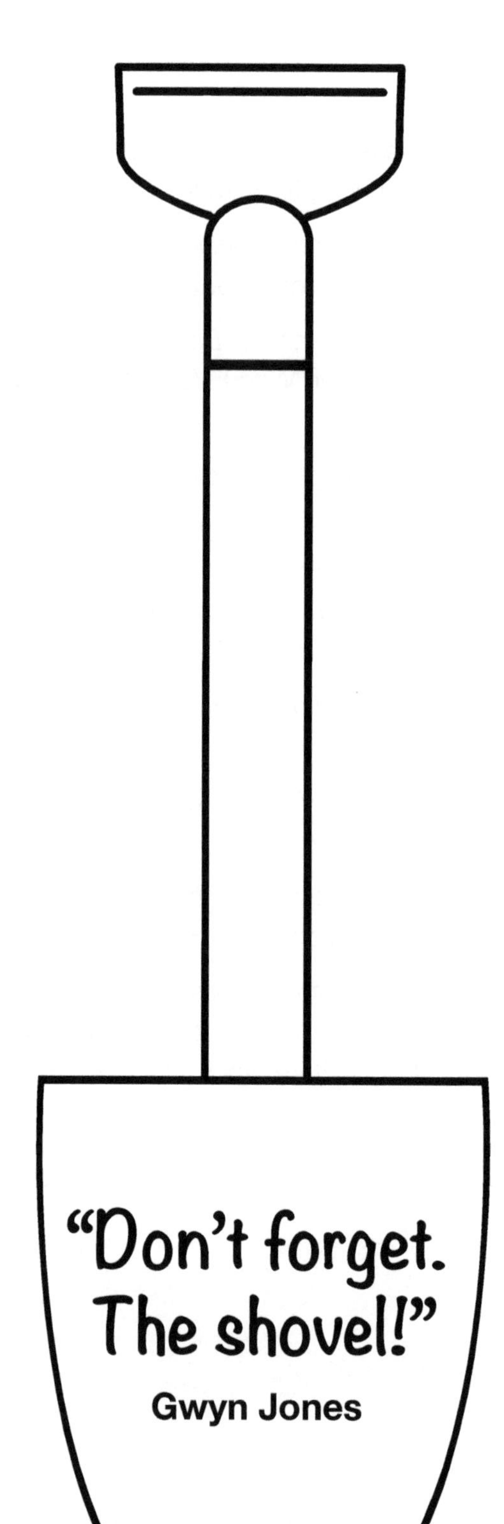
"Don't forget.
The shovel!"
Gwyn Jones

Printed in Australia
Ingram Content Group Australia Pty Ltd
AUHW011149171123
386657AU00002B/2

9 781925 370843